Microwave Semiconductor Devices

Microwave Semiconductor Devices

SITESH KUMAR ROY
Chief Investigator, Formerly Professor and Head
Institute of Radiophysics and Electronics
Calcutta University

MONOJIT MITRA
Department of Electronics and Telecommunication Engineering
Bengal Engineering College (Deemed University), Howrah

PHI Learning Private Limited
Delhi-110 092
2026

*In fond memory of **Shri Asoke K. Ghosh** (October 1942 – February 2024), Founder Chairman and Managing Director of PHI Learning, whose vision endlessly inspires.*

The Legacy Continues....

Published by Pushpita Ghosh, PHI Learning Private Limited, Rimjhim House, 111, Patparganj Industrial Estate, Delhi-110092 and Printed by Syndicate Binders, A-20, Hosiery Complex, Noida, Phase-II Extension, Noida-201305 (N.C.R. Delhi).

₹550.00

MICROWAVE SEMICONDUCTOR DEVICES
Sitesh Kumar Roy and Monojit Mitra

ISBN-978-81-203-2418-3 (Print Book)
ISBN-978-93-5443-688-8 (e-Book)

The export rights of the book are vested solely with the publisher.

Contents

Preface

Microwaves have been used extensively since the Second World War when the sources were based on vacuum devices. Microwaves (3–30 GHz) are presently playing a vital role in land based and satellite based communication and radars and have wide civilian and defence applications. In recent years, due to availability and emergence of semiconductor millimeter wave sources, millimeter waves (30–300 GHz) are finding wider application in ground and satellite communication, radars and missile guidance systems. This wide spectrum of application is making the millimeter wave system development one of the most advanced technology of radio science, specially in view of the ever increasing demand of communication and saturation of microwave frequency range with increasing number of channels. Millimeter wave semiconductor devices are presently under intensive investigation by various researchers all over the world although Sir J.C. Bose successfully generated electromagnetic waves of wavelength 6 mm and frequency 50 GHz in Calcutta as early as 1896.

This book is intended to provide a foundation to understand the physics and operational principle of microwave and millimeter wave semiconductor devices. These devices are making a revolutionary change in the field of communication and radar technology and will have important emerging applications in the twenty-first century.

The book deals with IMPATTs, Gunns, FETs, HEMTs etc., which are discussed in greater detail to provide fundamental knowledge to students, engineers and scientists. Various three-terminal and two-terminal devices in the microwave and millimeter wave field based on silicon and other Gr III-V semiconductors are discussed in the book.

The book will serve as a textbook for undergraduate electronics engineering students and a reference book for professional engineers and physicists.

We are indebted to the Centre of Advanced Study in Radio Physics and Electronics, University of Calcutta, for providing the facilities to write this book and also to Bengal Engineering College, Howrah, for necessary help to us. We are thankful to the co-investigators Dr. D. Ghoshal, Dr. N. Majumder, Sri B. Maity, Sri N.C. Mondal and Sri T.K. Pal.

Finally we thank Mr. Santanu Ghosh and Mr. Subhadeep Seal for typesetting, drawing figures and illustrations of the manuscript.

Sitesh Kumar Roy
Monojit Mitra

Chapter 1

Microwave and Millimeter Wave Devices

1.0 INTRODUCTION

Microwaves cover the wavelength range of 10–1 cm, and millimeter waves cover the wave length 1 cm–1 mm, corresponding to the high frequency electromagnetic oscillation ranging from 3 GHz–300 GHz. They have significant application in communication since they act as carriers of information in analog and digital form. Devices, which generate microwaves and millimeter waves, are essential for satellite and line of sight communication for audio and video signals as also for digital data communications. Moreover, microwaves are very useful for radars since they can penetrate through fog and cloud.

Millimeter waves are emerging very fast as highly suitable for carrying large number of channels of audio, video and data since the frequency is about 10 times more than microwaves and the microwave frequency-range as carriers of information is getting saturated for communication. Small antennas and sharp beams characterize millimeter wave systems. Millimeter wave devices are also being used in missile technology for defence applications.

The early generations of microwaves during the Second World War were vacuum devices, viz. Klystrons, Magnetrons and Travelling wave tubes, which depended on the motion of electrons in vacuum for various configurations of electric and magnetic fields. These tubes were heavy and bulky and required high voltage for their operation and occupied large space.

The invention of transistors led to almost complete replacement of vacuum triodes and pentodes in the audio and radio frequency range. The integration of a large number of transistors in very large-scale integrated circuits led to solid-state replacement on computer switching circuits, which were originally based on vacuum triodes and pentodes. Simultaneously, solid-state replacement of microwaves and millimeter waves started in the 1960s. Among these devices, Gunn diodes, IMPATT diodes, Bipolar transistors, Metal Semiconductor Field Effect Transistors (MESFET), and High Electron Mobility Transistors (HEMT) have emerged as highly suitable generators of microwaves and millimeter waves. IMPATT (IMPact ionization Avalanche Transit Time) covers the frequency range of 3–300 GHz, and is now the most powerful solid-state source covering the entire frequency spectrum of microwaves and millimeter waves. IMPATTs are based on p-n junctions of Ge, Si, GaAs and InP. Gunn diode covers the frequency range up to 100 GHz and provides sufficient low noise microwave power for communication. It is a bulk device based on n-GaAs or n-InP. Bipolar transistor, and MESFET operate at microwave frequency-range as ordinary bipolar transistor and MESFET with technological refinements. HEMT depends on formation of a two-dimensional electrons gas formed at the heterojunction of AlGaAs and GaAs when electrons move with very high speed under the gate electrode and generate oscillation up to 60 GHz. These semiconductor devices are basically small and reliable and they require low voltage for their operation. The oscillators and amplifiers based on these devices are finding wide application in satellite and terrestrial communications. Since computers worldwide are being linked through communication lines and satellites, microwave and millimeter wave devices will play a key role as microwave and millimeter wave sources in the expanding world of information technology.

A summary of energy band, carrier distribution, transport properties, and device technology including different methods for junction formation are presented emphasizing on the three most important semiconductors, germanium (Ge), silicon (Si), and gallium arsenide (GaAs). This will help to understand the device characteristics.

A detailed discussion is given on IMPATT devices, which have emerged as very powerful solid-state source of microwave, millimeter wave and submillimeter wave power. IMPATT is the acronym for IMPact Avalanche Transit Time and is basically a p-n junction diode of appropriate doping profile reversed biased to breakdown. Due to vast frequency range of operation and large power output IMPATT devices would find increasing applications in the 21st Century. Due to the importance of IMPATT devices in both microwave and millimeter wave range a special emphasis is given to it in this book. The physics of microwave and millimeter semiconductor devices, principle of operation, applications of microwave IMPATT devices and sources are therefore discussed in detail.

TRAPATT (TRApped Plasma Avalanche Triggered Transit) which is a solid-state microwave device with very high dc to microwave conversion efficiency and capable of producing high microwave power under pulsed mode operation is also discussed. Its frequency is restricted to lower microwave frequency range.

BARITT diodes (BARrier Injection Transit Time) is normally a p^+np^+ or metal–n–metal junction and is equivalent to a pair of diodes, connected back to back. BARITT can generate microwave power with very low noise but operates at lower microwave frequency.

Transfer electron devices, popularly called Gunn diodes, which depend for its operation on negative differential mobility in certain Gr-III & V semiconductors which arise due to the transfer of electrons from low to high valley conduction bands in n-type GaAs and n-type InP is discussed. These devices are bulk semiconductor devices based on band properties of Gr-III & V semiconductors. They are suitable as microwave and millimeter wave sources up to 100 GHz and are having large number of applications in microwave and millimeter wave systems.

Other types of junction diodes useful for microwaves and millimeter waves are tunnel diodes, backward diodes and Schottky diodes. Tunnel diodes are based on tunnelling of majority carrier across p-n junction. They have high speed and are used as pump sources in parametric amplifiers.

Metal semiconductor Schottky barrier diode is used as detector of microwave power. Under normal operating conditions it exhibits a square low response. The lower $1/f$ noise, better reliability and low burn out of this diode have made it more advantageous than the point contact diode.

High frequency bipolar as well as heterojunction transistors are also discussed. Most of the bipolar devices are fabricated from silicon and can give appreciable power up to 20 GHz.

MESFET is a field effect transistor having metal semiconductor junction, where the depletion layer is controlled by gate electric field. Its efficiency and frequency range is higher than those of a bipolar transistor. MESFET has emerged as a very important source of microwave of lower millimeter wave frequencies.

High Electron Mobility Transistor (HEMT), which has a selectively doped GaAs–AlGaAs heterojunction structure is discussed. HEMT is characterized by higher speed and higher frequency operation than MESFET. HEMT has a great promise and is a suitable device for VLSI chips, space communication and high-speed super computers.

Figure 1.1 shows the output power available at different frequencies from different devices.

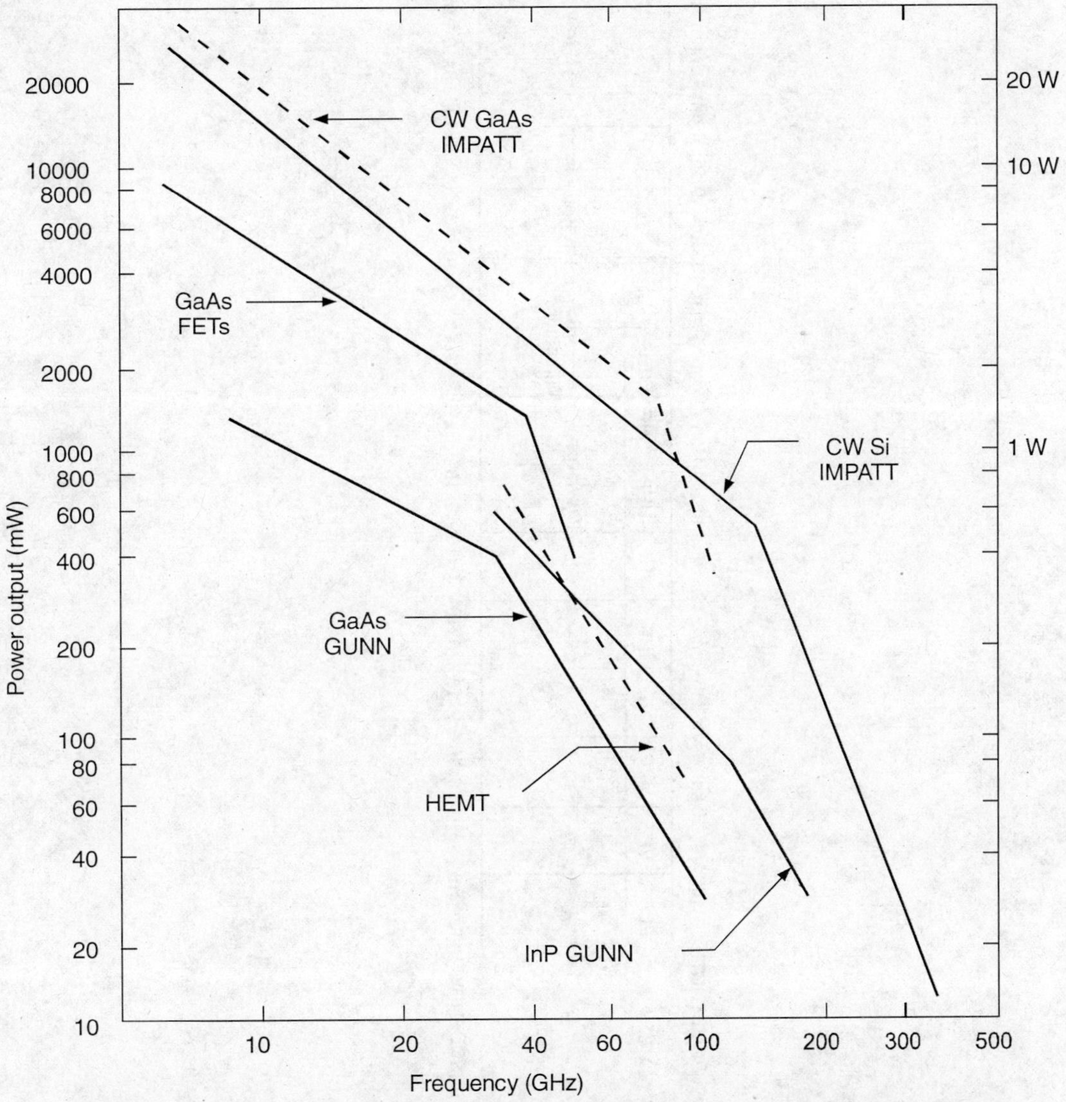

Figure 1.1 State of the art for microwave and millimeter wave solid state sources.

1.1 MICROWAVE REGIONS AND BAND DESIGNATION

Figure 1.2 shows the whole electromagnetic spectrum. Standard band designation for microwave frequencies listed as per the Institute of Electrical and Electronics Engineers (IEEE) in Table 1.1 is as follows:

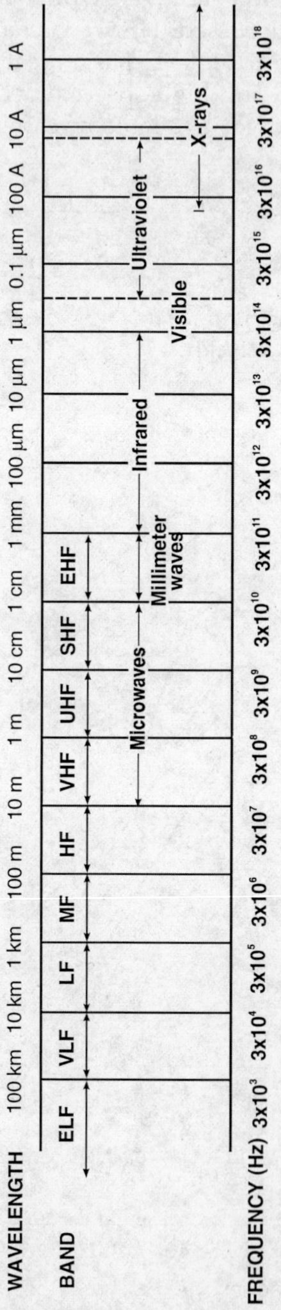

Figure 1.2 Electromagnetic spectrum.

Table 1.1

Designation	*Frequency range* (GHz)
HF	0.003 to 0.030
VHF	0.030 to 0.300
UHF	0.300 to 1.000
L-band	1.000 to 2.000
S-band	2.000 to 4.000
C-band	4.000 to 8.000
X-band	8.000 to 12.000
Ku-band	12.000 to 18.000
K-band	18.000 to 27.000
Ka-band	27.000 to 40.000
Millimeter	40.000 to 300.00
Sub-millimeter	300.000 and above

1.2 ADVANTAGE OF MICROWAVE AND MILLIMETER WAVE

(i) **High directivity:** Microwave and millimeter wave beams are highly directive and very sharp. With the increase of frequency beamwidth decreases and directivity increases. Relation between beamwidth (B) and directivity (D) is

$$B = 140\lambda/D$$

where λ is the wavelength of em wave.

(ii) **Improved bandwidth availability:** Large bandwidth (because of very high centre frequency, 3 GHz to few hundred GHz) of microwave and millimeter wave can accommodate more number of information channels for TV networks, telephone network, space communication, and defence etc. with a small percentage variation of carrier frequency.

(iii) **Fading effect and reliability:** At high frequency fading effect is less and hence microwave and millimeter wave communication is more reliable.

(iv) **Power requirements:** The power requirement for both transmitter and receiver for CW communication is low.

(v) **Transparency property of microwave:** Microwaves can freely propagate through different ionized layer as well as through the atmosphere. Millimeter waves can freely propagate easily at various window frequencies through the atmosphere.

Chapter 2

Physics of Semiconductor Devices

2.0 INTRODUCTION

Semiconductor is a class of material whose electrical conductivity lies between that of a conductor and an insulator. At room temperature the electrical conductivity of a conductor such as a metal is about 10^8 siemens per metre whereas that of an insulator like diamond is about 10^{-10} siemens per metre. The conductivity of a semiconductor lies between 10^5–10^{-4} siemens per metre. It is generally sensitive to temperature, magnetic field, illumination, and the amount of impurity atoms, which may vary from 1 μg to 1 g in 1 kg of semiconductor material. A pure semiconductor has a negative temperature coefficient of resistance but at room temperature all semiconductors are insulators. The occupancy of electrons in the highest band or in the two higher bands determine the conductivity of a solid. Figure 2.1 shows the energy band diagrams of three classes of solid-metals, semiconductors, and insulators.

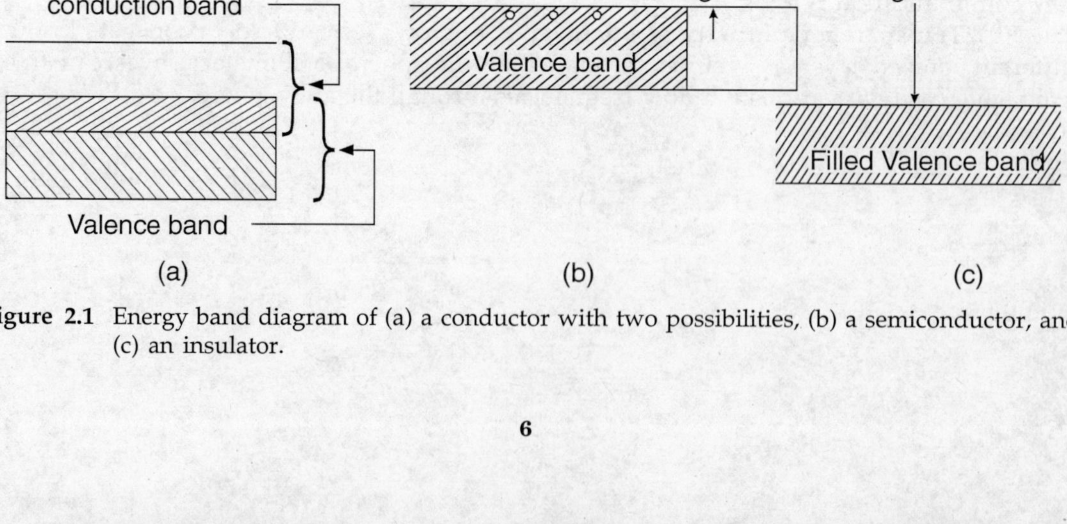

Figure 2.1 Energy band diagram of (a) a conductor with two possibilities, (b) a semiconductor, and (c) an insulator.

Metals

In a conductor such as metal the conduction band either is partially filled (copper) or overlaps the valence band (such as zinc or lead) so that there is no bondage as shown in Figure 2.1(a). Here the electrons form the top of the valence band or the upper most electrons in the partially filled band can move easily to the next higher level when they gain kinetic energy due to the applied field. Due to the presence of so many unoccupied states near the occupied energy state, electrons can easily move under a small magnitude of applied field. Thus conduction of current is easily possible in a metal.

Insulators

In an insulator the valence electrons are strongly coupled under the bonds and, as a result no free electrons are available in the conduction band, hence conduction is not possible. As shown in the Figure 2.1(c), insulator possesses a large bandgap and no energy is sufficient to bring an electron from valence to the conduction band. Ceramic, polyester, silicon dioxide are some examples of the insulators.

Semiconductors

In semiconductors no free electrons are available in the conduction band at 0 K but bandgap is much smaller than that of an insulator. Thus, because of many available empty states in the conduction band, the electrons can easily jump from valence band to the conduction band with gain of energy from the applied field and can take part in current carrying mechanism. The bandgap of Si is 1.12 eV and that of GaAs is 1.42 eV. The energy band diagram of a semiconductor is as shown in Figure 2.1(b).

2.1 SEMICONDUCTOR MATERIALS

The study of semiconductor materials began in the early nineteenth century [1] and till date many semiconductors have been investigated. Semiconductors are of two types—(a) element, and (b) compound semiconductors.

2.1.1 Element semiconductor

They are mainly the elements of Gr–IV of the periodic table such as silicon (Si) and germanium (Ge). In the early 1950s, germanium was the major semiconductor material. From 1960 silicon replaced germanium completely because of its low cost, simple metallurgical technology, easy availability, and as it shows better semiconductor property at room temperature. SiO_2 is a very stable compound and can be used as a mask in integrated circuit fabrication.

2.1.2 Compound semiconductor

Combining two elements of either Gr(III-V) or of Group (II–VI) of the periodic table this type of semiconductor is made. Compound semiconductors like GaAs, GaP, InP, InAs can be combined to yield ternary or quaternary compound semiconductor like $Ga_xIn_{1-x}As_yP_{1-y}$. These compound semiconductors have electrical and optical properties that are different from silicon. These semiconductors, especially GaAs are used mainly for high-speed electronic and photonic applications. These devices fabricated from compound semiconductors have higher efficiencies, but the technology is much more complicated than that of silicon.

2.2 CLASSIFICATION OF SEMICONDUCTORS

Semiconductors are basically of two types—intrinsic and extrinsic. The conduction in the intrinsic semiconductor depends entirely on the thermally generated valence electrons; it is known as a pure semiconductor. Silicon and germanium are the two important elemental semiconductors. A two dimensional picture of a silicon crystal is as shown in Figure 2.2. Here four electrons in the outermost shell of the silicon atom form the covalent bond with the neighbouring silicon atoms. This electron pair or covalent bond is denoted in Figure 2.2 by two dashed lines joining each atom to each of the adjacent atoms. Since the valence electrons bind one atom to another, they are not available for electrical conduction in the absence of thermal excitation. As mentioned earlier in the energy band diagram of Figure 2.1 at 0 K the valence band is completely empty.

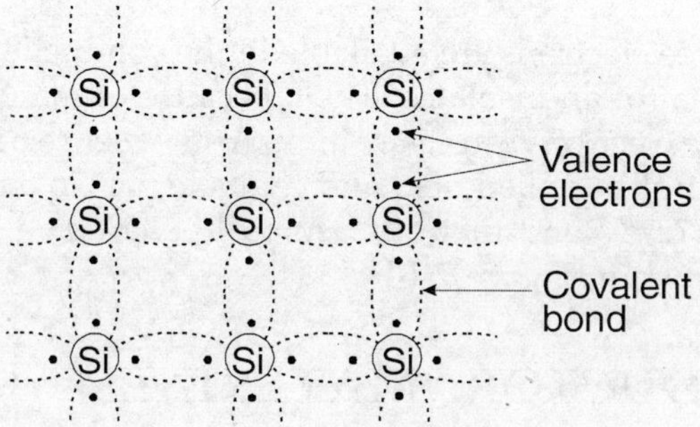

Figure 2.2 Structure of silicon crystal.

At room temperature some of the valence electrons have enough thermal energy (KT_0) to break the valence band and become free to move randomly throughout the crystal. At room temperature silicon requires 1.12 eV of energy to break the covalent bond, which is the forbidden energy gap. The vacancy thus created by this process in the covalent bond is known as hole; and as it is created due to the shortage of electron, it is of positive charge. The holes also contribute to the conductivity of the semiconductor. When the silicon atom is thermally excited it results in the creation of electron-hole pair. So in an intrinsic semiconductor the number of holes is equal to the number of electrons. Again as holes and electrons are oppositely charged, they attract each other and recombine. So some electron-hole pairs are lost due to the recombination. But at equilibrium the electron concentration n and hole concentration p are equal; i.e., $n = p = n_i$, where n_i is termed as the intrinsic carrier concentration. When an electron-hole pair is thermally created, a valence electron in a neighbouring atom can have sufficient thermal energy to jump into the position of the hole and reconstruct the covalent bond and in the process the electron leaves a hole in its initial position. Thus the holes move in the direction opposite to those of valence electrons. The movement of an electron and a hole in a semiconductor is as shown in Figure 2.3.

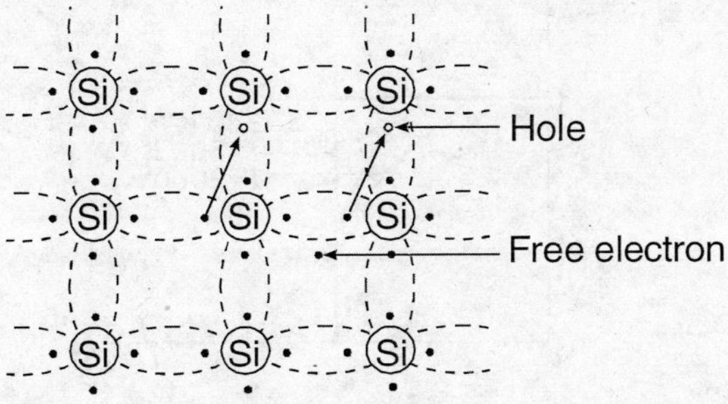

Figure 2.3 Movement of a hole in a semiconductor.

2.2.1 Intrinsic carrier concentration

When a thermally excited electron jumps from valence band to conduction band it leaves equal number of holes in the valence band. To obtain the electron density in an intrinsic semiconductor, one has to calculate the electron density in an incremental energy range dE. This density $n(E)$ is given by the product of the density of states $N(E)$, that is the density of allowed energy states per energy range per unit volume and by the probability of occupying that energy range $F(E)$. Thus integrate $N(E)\ F(E)\ dE$ in between the limit 0 and E_{Top} i.e., from the bottom of the conduction band to E_{Top}, which is the top of the conduction band.

$$n = \int_0^{E_{\text{Top}}} n(E)\, dE = \int_0^{E_{\text{Top}}} N(E)F(E)\, dE \tag{2.1}$$

Now Fermi-Dirac distribution function gives the probability that an electron occupies an electronic state having energy E

$$F(E) = \frac{1}{1 + \exp^{(E-E_F)/KT}} \tag{2.2}$$

where

K = Boltzmann constant

T = Absolute temperature

E_F = Fermi energy level

At absolute zero of temperature, i.e., at $T = 0$ K, Equation (2.2) shows that $F(E) = 1$ for $E < E_F$ and $F(E) = 0$ for $E > E_F$. Thus the electron occupies all energy states below E_F and all the energy states above E_F are completely empty. Hence the Fermi energy E_F denotes the maximum energy that can be occupied by electrons at 0 K. When $E = E_F$ for $T > 0$, Equation (2.2) shows that $F(E) = 1/2$ which signifies that the Fermi level is the energy level, for which the probability of occupancy is 1/2 for a finite non-zero temperature. This is explained with the help of Fermi distribution function $F(E)$ versus (E/E_F) for various temperature graphs of Figure 2.4. In order to find out the value of $n(E)$ of Equation (2.1), we will take the help of Figure 2.5 (a), (b), (c) and (d) where $n(E)$ and $F(E)$ are plotted in terms of energy E.

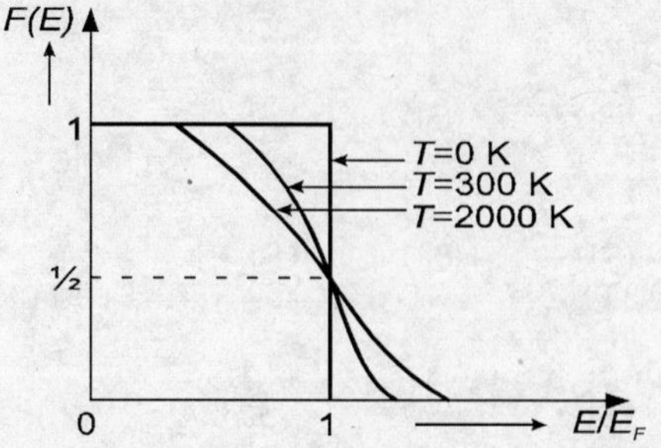

Figure 2.4 Fermi distribution function $F(E)$ versus E/E_F plot for various temperatures.

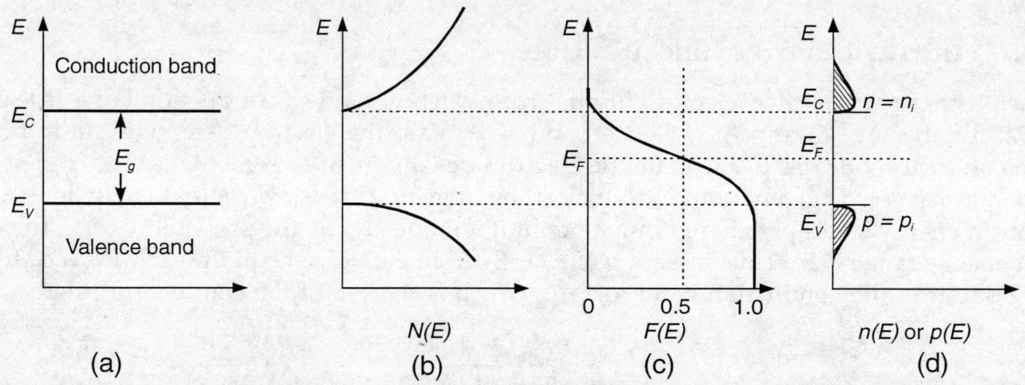

Figure 2.5 Intrinsic semiconductor: (a) schematic band diagram, (b) density of states, (c) Fermi distribution function, and (d) carrier concentration.

Now the Fermi distribution function of Equation (2.2) can then be approximated as

$$F(E) \cong e^{-(E-E_F)/KT} \text{ for } (E - E_F) > 3\,KT \tag{2.3}$$

Substituting the approximation of Equation (2.3) and the values from the grophs of Figure. 2.5 in Equation (2.1) we get

$$n = \frac{2}{\sqrt{\pi}} N_c (KT)^{-3/2} \int_0^{\alpha} E^{1/2} \exp\left[-(E-E_F)/KT\right] dE \tag{2.4}$$

where

$$N_c = 12(2\pi m_\alpha\, KT/h^2)^{3/2} \text{ for silicon} \tag{2.5a}$$

$$= 12(2\pi m_\alpha\, KT/h^2)^{3/2} \text{ for GaAs} \tag{2.5b}$$

Let $E/KT = x$ then Equation (2.4) becomes

$$n = \frac{2}{\sqrt{\pi}} N_c \exp(E_F/KT) \int_0^{\alpha} x^{1/2} e^{-x} dx \tag{2.6}$$

$$= \frac{2}{\sqrt{\pi}} N_c \exp(E_F/KT)\ \frac{\sqrt{\pi}}{2}$$

Therefore,

$$n = N_c \exp\,(E_F/KT) \tag{2.7}$$

Now if the bottom of the energy band is E_c instead of $E = 0$, then

$$n = N_c \exp\,[-(E_c - E_F)/KT] \tag{2.8}$$

Similarly the hole density

$$p = N_v \exp\,[-(E_F - E_v)/KT] \tag{2.9}$$

where

$$N_v = 2(2\pi m_p\, KT/h^2)^{3/2} \tag{2.10}$$

$$= \text{effective density of state in the valence band}$$

or in an intrinsic semiconductor the number of electrons/unit volume in the conduction band is equal to that of holes in the valence band. The Fermi level for an intrinsic semiconductor is obtained by equating Equations (2.8) and (2.9)

$$E_F = E_c = (E_c + E_v)/2 + (KT/2)\ \ln\,(N_v/N_c) \tag{2.11}$$

At room temperature the second term is much smaller than the band gap. Hence Equation (2.11) reduces to

$$E_c = E_F = (E_c + E_v)/2$$

Thus, the Fermi level for an intrinsic semiconductor lies very close to the bandgap. The intrinsic carrier density is obtained from Equations (2.8), (2.9) and (2.11).

$$\begin{aligned} N_i^2 &= pn \\ &= N_c N_v \exp\,[-(E_c - E_F)/KT - (E_F - E_v)/KT] \\ &= N_c N_v \exp\,[-(E_c - E_v)/KT] \\ &= N_c N_v \exp\,[-E_g/2KT] \end{aligned} \tag{2.12}$$

2.2.2 Extrinsic semiconductors

When a small amount of impurity atoms of either Gr-III or of Gr-V is added to the intrinsic semiconductor, free holes or electrons are generated by excitation of the impurity atoms. The process is called doping and the impurity that is added is known as dopant, and the semiconductor after doping is called an extrinsic semiconductor. When a tetravalent semiconductor (silicon) is doped with a pentavalent element like arsenic or phosphorus, a silicon atom is replaced by arsenic with five valence electrons. The arsenic atom forms covalent bonds with its four neighbouring silicon atoms. The fifth electron has a relatively small bonding energy than its host arsenic atom and can be ionized to become a conduction electron at a moderate temperature. Thus it can be said that the arsenic atom donates an electron to the conduction band. The arsenic atom is called the donor atom and the silicon crystal becomes the n-type semiconductor. The schematic band pictures for n-type silicon with donor (arsenic) atom is shown in Figure 2.6. Similarly, when trivalent impurity elements like boron or aluminium substitutes a silicon atom, an additional electron is accepted to form four covalent bonds and a deficiency of electron thus creates a positively charged hole in the semiconductor. This is p-type semiconductor and the boron is an acceptor atom. In Figure 2.7(a) and (b) the energy band diagram of an extrinsic semiconductor with donor and acceptor ions are shown.

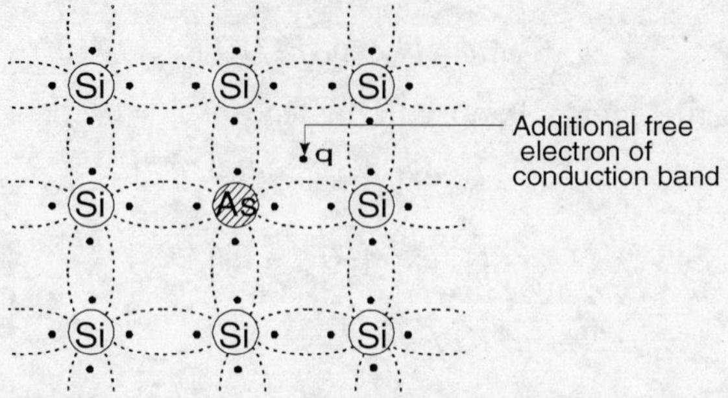

Figure 2.6 n-type Si with donor (arsenic) impurity.

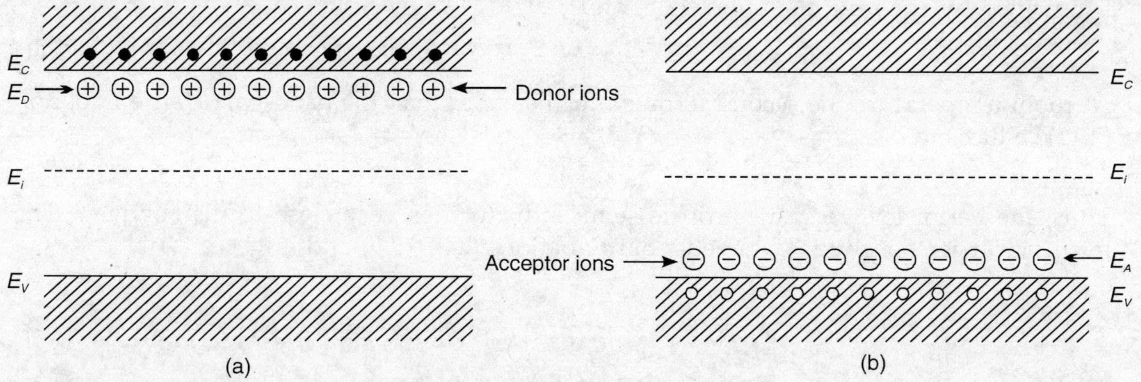

Figure 2.7 Energy band diagram of extrinsic semiconductor with (a) donor ions and (b) acceptor ions.

2.2.3 Non-degenerate semiconductors

If the electron or hole concentration is much lower than the effective density of states in the conduction band, the semiconductor is referred to as non-degenerate semiconductor. Under complete ionization condition, the electron density

$$n = N_D \tag{2.13}$$

where N_D is the donor concentration.

From Equations (2.8) and (2.13), we can obtain the Fermi level in terms of the effective density of states N_c and the donor concentration N_D.

$$N_D = N_c \exp [-(E_c - E_F)/KT]$$

or

$$E_c - E_F = KT \ln (N_c/N_D) \tag{2.14}$$

Similarly for holes,

$$E_F - E_v = KT \ln (N_v/N_A) \tag{2.15}$$

Now

$$n = N_c \exp[-(E_c - E_F)/KT]$$
$$= N_c \exp[-(E_c - E_i)/KT] \exp[(E_F - E_c)/KT]$$
$$= n_i \exp[(E_F - E_i)/KT] \quad (2.16a)$$

and

$$p = n_i \exp[(E_i - E_F)/KT] \quad (2.16b)$$

where E_i = Intrinsic Fermi level.

For an extrinsic semiconductor either electron or hole will dominate but the product of two types of the carriers will remain constant at a given temperature. The neutrality condition is

$$n + N_A = p + N_D \quad (2.17)$$

Solving Equation (2.17) with $n \cdot p = n_i^2$, we get the equilibrium electron or hole concentration in an n-type semiconductor

$$n_n = \frac{1}{2}\left[N_D - N_A + \sqrt{(N_D - N_A)^2 + 4n_i^2}\right] \quad (2.18)$$

$$p_n = n_i^2/n_n \quad (2.19)$$

Similarly, concentration of holes and electrons in a p-type semiconductor are given by

$$P_p = \frac{1}{2}\left[N_D - N_A + \sqrt{(N_D - N_A)^2 + 4n_i^2}\right] \quad (2.20)$$

$$n_p = n_i^2/p_p \quad (2.21)$$

The subscript n and p represent n- and p-type semiconductor respectively. Generally, the magnitude of the net impurity concentration $|N_D - N_A|$ is larger than the intrinsic carrier concentration n_i. The above relationship can be simplified to

$$N_n = N_D - N_A \qquad \text{when } N_D > N_A \quad (2.22a)$$

$$P_p = N_A - N_D \qquad \text{when } N_A > N_D \quad (2.22b)$$

As the temperature of the semiconductor rises, we have seen that both n and p-type semiconductors have become essentially intrinsic. Thus, when the temperature of n-type semiconductor increases, the Fermi level (E_F) moves towards the centre of the bandgap. The variation of the Fermi level with temperature and impurity concentration is depicted in Figure 2.8.

2.2.4 Degenerate semiconductors

When the doping concentration of a semiconductor becomes equal or larger than the corresponding effective density of states, the semiconductor is referred to as a degenerate semiconductor. For a heavily doped n-type or p-type semiconductor, E_F will be above E_c or below E_v and this high impurity concentration causes a reduction of the bandgap. The bandgap reduction ΔE_g for silicon at room temperature is given by

$$\Delta E_g = 22 \times 10^{-9} \sqrt{N}$$

where the doping is in cm^{-3}.

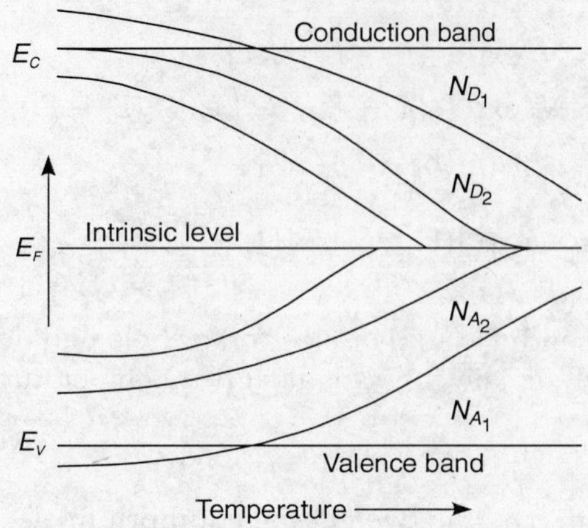

Figure 2.8 Variation of Fermi level in a semiconductor with temperature and impurity concentration.

2.3 CARRIER MOBILITY

Behaviour of electrons in a solid for the sake of simple mathematical analysis can be considered analogous to the movement of gas molecules inside a closed chamber. In that case electrons may be treated as particles of dynamics. Consider that an electron with an initial velocity of zero collides with another electron in course of its random motion inside a solid after the expiry of a very small time known as mean free time (τ_e). If the electric field applied to the solid is E, then the acceleration acquired by the electron is

$$f_e = \frac{\text{Force}}{\text{Effective mass}} = -\frac{qE}{m} \tag{2.23}$$

Therefore the velocity after a time τ_e is given by

$$v_e = f_e \tau_e = \frac{-qE}{m} \cdot \tau_e \tag{2.24}$$

The average velocity due to the application of electric field is termed as drift velocity and is given by

$$v_d = (0 + v_e)/2 = \frac{-qE\tau_e}{2m} \tag{2.25}$$

Here the negative sign indicates that the velocity of electron is opposite to that of applied electric field.

Mobility of a carrier can now be defined as the drift velocity per unit applied electric field.

Therefore,

$$\mu = \frac{v_d}{E} = \frac{q\tau_e}{2m} \tag{2.26}$$

From Equation (2.26) it is clear that the mobility of an electron is much more than that of a hole because of the lower effective mass of electron than the hole.

2.4 ELECTRICAL CONDUCTIVITY

Let us consider a cylinder of semiconductor material having cross-sectional area α, through which the charge is moving with drift velocity v_d from one end to the other. In one second the carrier can move a distance v_d, and let us consider the cylinder of length v_d. If n_0 is the carrier concentration, then the number of carriers within the cylinder is $n_0 \; \alpha v_d \; q$, where q is the charge of a carrier. As this charge crosses the length v_d of the cylinder per second, the electric current flowing through the semiconductor is given by

$$I = n_0 \alpha v_d q \tag{2.27}$$

The current density is

$$J = \frac{I}{\alpha} = n_0 q v_d \text{ A/m}^2 \tag{2.28}$$

Now from Equation (2.26), $v_d = \mu E$; and thus Equation (2.28) can be written as

$$J = n_0 q \mu E \tag{2.29}$$

The electrical conductivity is the current density per unit electrical field

$$\sigma = \frac{J}{E} = n_0 q \mu \tag{2.30}$$

When current carriers are electrons, the relevant conductivity of electrons is $\sigma_n = nq\mu_n$, and the conductivity for the holes is $\sigma_p = pq\mu_p$, where n and p are the electron and hole concentrations respectively. In a semiconductor both the types of carriers are present and thus the overall conductivity is

$$\sigma = \sigma_n + \sigma_p = q(n\mu_n + p\mu_p) \tag{2.31}$$

For an intrinsic semiconductor $n = p = n_i$ and thus Equation (2.31) becomes

$$\sigma_i = q n_i (\mu_n + \mu_p) \tag{2.32}$$

Problem 2.1 Determine the conductivity and resistivity at 300 K of pure germanium, assuming that at 300 K the concentration of carriers is 2.5×10^{13} cm^{-3} and carrier mobilities 3600 cm^2/V.s for the electrons and 1700 cm^2/V·s for the holes.

Solution: Given data

$$\mu_n = 3600 \text{ cm}^2/\text{V·s}$$

$$\mu_p = 1700 \text{ cm}^2/\text{V·s}$$

$$n_i = 2.5 \times 10^{13} \text{ cm}^{-3}$$

Therefore, the conductivity of germanium at 300 K is given by

$$\sigma_i = n_i(\mu_n + \mu_p)q = 2.5 \times 10^{13}(3600 + 1700) \times 1.6 \times 10^{-19}$$

$$= 0.0212 \text{ mho/cm}$$

and resistivity

$$\rho_i = \frac{1}{\sigma_i} = \frac{1}{0.0212} = 47.17 \; \Omega \text{ cm}$$

Problem 2.2 (a) Show that the minimum conductivity of a semiconductor sample occurs when

$$n_0 = n_i\sqrt{\mu_p/\mu_n}$$

(b) What is the expression for minimum conductivity σ_{min}?

Soultion: (a) Conductivity of a semiconductor is given by

$$\sigma = q(n_0\,\mu_n + p_0\,\mu_p) \tag{2.33}$$

where n_0 and p_0 are the electron and hole concentrations at equilibrium. Now the intrinsic carrier concentration (n_i)

$$n_i^2 = n_0 p_0 \tag{2.34}$$

Substituting the value of p_0 from Equation (2.34) into Equation (2.33)

$$\sigma = q\left[n_0\mu_n + (n_i^2/n_0)\mu_p\right] \tag{2.35}$$

Minimum conductivity means when $d\sigma/dn_0 = 0$.

Thus

$$q\,[\mu_n - \mu_p\,n_0^{-2}\,n_i^2] = 0$$

Therefore

$$\mu_n = \mu_p\,\frac{n_i^2}{n_0^2}$$

or

$$n_0 = n_i\sqrt{\mu_p/\mu_n} \tag{2.36a}$$

(b) Substituting the value of n_0 from Equation (2.36a) to Equation (2.35), we get

$$\sigma_{min} = q\left[n_i\sqrt{\frac{\mu_p}{\mu_n}}\,\mu_n + \frac{n_i^2}{n_i\sqrt{\mu_p/\mu_n}}\,\mu_p\right]$$

$$= 2qn_i\sqrt{\mu_p\mu_n} \tag{2.36b}$$

Problem 2.3 A specimen of silicon has square cross section of 2×2 cm^2 and length of 2 cm. The current is due to electrons whose mobility is 1300 cm^2/V·s. An applied dc voltage of 1 V across the bar produces a current of 8 mA in it.

Calculate:

(a) Concentration of free electron

(b) Drift velocity

Solution: (a) The cross sectional area A of the bar is

$$A = 2 \times 10^{-2} \times 2 \times 10^{-2} = 4 \times 10^{-4}\ \text{m}^2$$

mobility μ_n = 1300 cm^2/V·s = 0.13 m^2/V·s

Using Ohm's law the resistance of the bar is

$$V/I = 1/8 \times 10^{-3} = 125\ \Omega$$

Therefore,

$$\text{Conductance} = 1/125$$

and conductivity

$$\sigma_n = \frac{1}{\rho} = \frac{l}{A} \cdot \frac{1}{125} = \frac{2 \times 10^{-2}}{4 \times 10^{-4}} \cdot \frac{1}{125} = 40\ (\Omega \cdot \text{m})^{-1}$$

Again $\sigma_n = q\ n\ \mu_n$.
Therefore,

$$n = \frac{\sigma_n}{q \cdot \mu_n} = \frac{40}{1.6 \times 10^{-19} \times 0.13} = 1.923 \times 10^{21}\ \text{atoms/m}^3$$

(b) Drift velocity $v_d = \mu\ E$

Here

$$E = \frac{V}{l} = \frac{1}{2 \times 10^{-2}} = 50\ \text{V/m}$$

Therefore, Drift velocity $v_d = 0.13 \times 50 = 6.5$ m/s.

Problem 2.4 A sample of Ge is doped with 10^{14} atoms of acceptor impurity and 10^{13} atoms of donor impurity per cc. Calculate the conductivity of the extrinsic material. Will this material be useful for semiconductor devices? Why?

Given:

$$n_i = 2.5 \times 10^{13}\ \text{cm}^{-3}$$
$$\text{atomic density} = 4.4 \times 10^{22}\ \text{cm}^{-3}$$
$$\mu_n = 0.38\ \text{m}^2/\text{V·s}$$
$$\mu_p = 0.18\ \text{m}^2/\text{V·s}$$
$$\text{and intrinsic resistivity} = 45\ \Omega\text{·cm}$$

Solution: In this sample the doping has been done by both donor impurity and acceptor impurity. Hence there would be sufficient number of holes and electrons. To start with the number of holes $p = N_A = 10^{14}\ \text{cm}^{-3}$ and the number of electrons $n = N_D = 10^{13}\ \text{cm}^{-3}$. Numbers of electrons and holes will be reduced due to recombination. We now calculate the final concentration of holes and electrons using the relationship

$$np = n^2$$

$$p - n = N_A - N_D = 10^{14} - 10^{13} = 0.9 \times 10^{14}$$

or

$$p = n + 0.9 \times 10^{14}$$

Since

$$np = n_i^2$$

Therefore

$$n(n + 0.9 \times 10^{14}) = (2.5 \times 10^{13})^2$$

or

$$n^2 + 0.9 \times 10^{14} n - 6.25 \times 10^{26} = 0$$

Solving the quadratic equation we get

$$n = 6.5 \times 10^{12} \text{ electrons/cm}^3$$

and

$$p = n + 0.9 \times 10^{14} = 6.5 \times 10^{12} + 0.9 \times 10^{14}$$
$$= 0.965 \times 10^{14} \text{ holes/cm}^3$$

Therefore, conductivity

$$\sigma = (n\,\mu_n + p\mu_p)q$$
$$= (6.5 \times 10^{12} \times 3000 + 0.965 \times 10^{14} \times 1800) \times 1.602 \times 10^{-19}$$
$$= 0.03174\ (\Omega\cdot\text{cm})^{-1}$$

Conductivity of intrinsic Ge = 1/45 = 0.0222$(\Omega\cdot\text{cm})^{-1}$.

Since the conductivity of the extrinsic material (0.03174) is higher than that of the intrinsic material (0.0222), it is usable as semiconductor. Also, since $p >> n$, the extrinsic semiconductor is effectively p-type.

2.5 EFFECTS OF TEMPERATURE AND DOPING ON MOBILITY

Two basic types of scattering mechanism that influence carrier mobility are (i) lattice scattering and (ii) impurity scattering. The former occurs due to the vibration of the lattice resulting from the temperature of the crystal. It is expected that the mobility should decrease as the sample is heated. This is indicated by the approximate $T^{-3/2}$ dependence when lattice scattering dominates as shown in Figure 2.9. On the other hand scattering from crystal defects such as ionized impurities become dominant as the temperature is lowered. Since a slowly moving carrier is likely to be scattered more strongly by an interaction with a charged ion then a carrier with greater momentum, impurity scattering events depict a decrease in mobility with decrease in temperature ($T^{3/2}$ dependence). In fact, if μ_l and μ_i denotes mobility values for lattice scattering and impurity scattering, the resulting mobility is given by

$$\mu = \frac{\mu_i \mu_l}{\mu_i + \mu_l} \tag{2.37}$$

With the increase of doping concentration the mobility decreases.

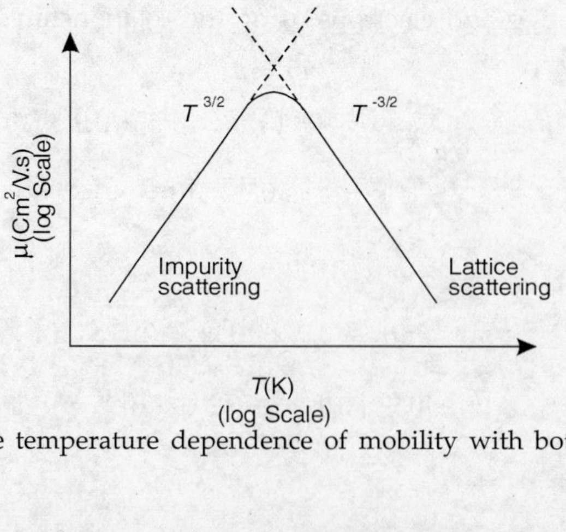

Figure 2.9 Approximate temperature dependence of mobility with both lattice and impurity.

2.6 CARRIER DIFFUSION

Because of the concentration gradient of charge carriers within a semiconductor, the charge particles tend to move from a region of higher concentration to a region of lower concentration in order to destroy the non-uniformity of the concentration. This phenomenon is known as diffusion.

Thus the diffusion current density J_n for electrons is proportional to the concentration gradient, i.e.

$$J_n \propto \frac{dn}{dx}$$

or

$$J_n = qD_n \frac{dn}{dx} \tag{2.38}$$

where D_n is the diffusion constant for electrons.

A similar expression can be derived for diffusion current density J_p for holes, where

$$J_p = -qD_p \frac{dp}{dx} \tag{2.39}$$

The negative sign in the above equation arises because dp/dx is negative when the charge density falls with increase in x and D_p (the diffusion constant). Equations (2.38) and (2.39) are also referred to as Fick's law.

Again when a potential gradient is created in a semiconductor due to the applied voltage, the charge carrier will gain energy from the applied field and moves with corresponding drift velocity. This current is referred to as drift current. Thus two components of drift currents are

$$\text{due to electrons} = qn\mu_n E$$

and

$$\text{due to holes} = qp\mu_p E \tag{2.40}$$

Hence the total electron current due to drift and diffusion is

$$J_n = qn\mu_n E + qD_n \frac{dn}{dx} \tag{2.41}$$

The total hole current due to drift and diffusion is

$$J_p = qp\mu_p E - qD_p \frac{dp}{dx} \tag{2.42}$$

The total current due to holes and electrons is the sum of the above two currents i.e.,

$$J = J_n + J_p \tag{2.43}$$

2.7 EINSTEIN'S RELATIONSHIP

The diffusivity D and the mobility μ are related to each other through the following relationship, called Einstein's equation and is given by

$$\frac{D_p}{\mu_p} = \frac{D_n}{\mu_n} = \frac{kT}{q} \tag{2.44}$$

where k is the Boltzmann constant, T is the absolute temperature and q is the electronic charge. At room temperature (300 K), putting $k = 1.38 \times 10^{-23}$ J and $q = 1.6 \times 10^{-19}$ coulomb, we obtain from the Equation (2.44)

$$\frac{D_p}{\mu_p} = \frac{D_n}{\mu_n} = 0.026 \text{ V} \tag{2.45}$$

At 300 K for electrons in Ge, the mobility $\mu_n = 0.38$ m^2/V·s and thus from Equation (2.45) we get $D_n = 0.026 \times 0.38 = 0.0099$ m^2/s.

2.8 LIFETIME OF MINORITY CARRIERS

At thermal equilibrium $pn = n_i^2$; but due to the injection of excess carrier $np > n_i^2$ and as a result the thermal equilibrium condition is disturbed. The number of these excess carriers decays with time through the process called recombination. This recombination phenomenon can be classified as direct and indirect process. Direct recombination is also called band-to-band recombination, which usually dominates in direct bandgap semiconductor like GaAs, InP etc. On the otherhand indirect recombination or so called via bandgap recombination dominates in indirect bandgap semiconductors like silicon, germanium etc.

At thermal equilibrium let the concentration of free electrons and holes be n_0 and p_0 respectively. If Δn and Δp be the small changes of electron and hole concentration due to injection, then the recombination rate is given by

$$R' = K(n_0 + \Delta n)\,(p_0 + \Delta p) \tag{2.46}$$

Where K is the proportionality constant. At thermal equilibrium $\Delta n = \Delta p = 0$ and rate of generation is equal to the rate of recombination. Thus the recombination rate for excess carrier is

$$\begin{aligned} R &= K[(n_0 + \Delta n)(p_0 + \Delta p) - n_0 p_0] \\ &= K[n_0 \Delta p + p_0 \Delta n + \Delta n \cdot \Delta p] \end{aligned} \tag{2.47}$$

As Δn and Δp are very small quantities $\Delta n \cdot \Delta p$ can be neglected from Equation (2.47).

Hence

$$R = K[n_0\, \Delta p + p_0\, \Delta n] \tag{2.48}$$

Now depending on the type of doping, one of the terms inside the bracket will be greater than the other; for example, in a p-type semiconductor $p_0\, \Delta n >> n_0\, \Delta p$; since $p_0 >> n_0$ and $\Delta n = \Delta p$.

Therefore Equation (2.48) reduces to

$$R = K p_0 \Delta n \quad \text{for p-type} \tag{2.49a}$$

$$R = K n_0 \Delta p \quad \text{for n-type} \tag{2.49b}$$

Thus from Equations (2.49a) and (2.49b) it is clear that the recombination rate is proportional to the concentration of excess minority carriers, which are electrons for a p-type semiconductor and holes for an n-type semiconductor.

For a p-type semiconductor Equation (2.49a) can be written as:

$$R = -\frac{d}{dt} \cdot (\Delta n) = \frac{\Delta n}{\tau_n} \tag{2.50}$$

Where τ_n is known as lifetime of excess minority carriers and is equal to $1/Kp_0$. In the same fashion one can write the recombination rate for excess minority carrier for an n-type semiconductor.

$$R = -\frac{d}{dt}\cdot(\Delta p) = \frac{\Delta p}{\tau_p} \tag{2.51}$$

2.9 CONTINUITY EQUATION

So far we have seen that in a semiconductor the drift of charge is due to the field; diffusion is due to concentration gradient, and recombination is due to the presence of charge of opposite polarity. These three effects are simultaneously present in a semiconductor; and the equation, which governs all these three effects, is known as the continuity equation.

An infinitesimal small thickness *dx* of a uniform semiconductor bar of cross-section *A* of Figure 2.10 is considered to derive the one dimensional continuity equation for electrons.

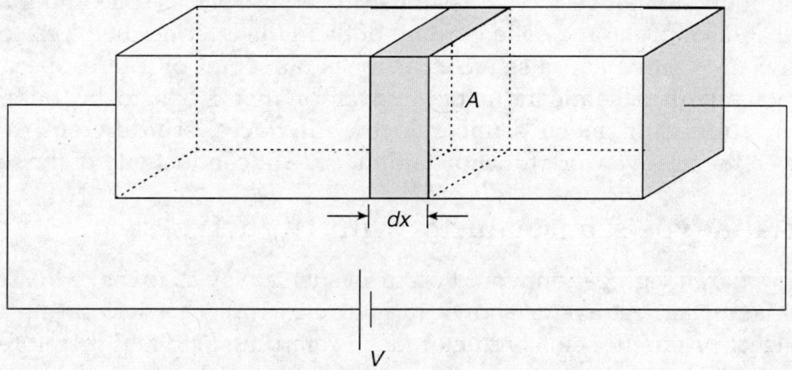

Figure 2.10 Structure of a uniform silicon bar.

Let $J_n(x)$ and $J_n(x + dx)$ be the current densities at the two sides of the infinitesimal small thickness *dx* of the bar. Also we are considering that G_n and R_n are the generation and recombination rates respectively within the region *dx* of the bar. Thus the overall rate of change in the number of electrons in the slice *dx* is:

$$\frac{\partial n}{\partial t} A\,dx = \left[\frac{J_n(x)A}{-q} - \frac{J_n(x+dx)\cdot A}{-q}\right] + (G_n - R_n)A\,dx \tag{2.52}$$

The Taylor series yields

$$J_n(x+dx) = J_n(x) + \frac{\partial J_n}{\partial t}dx \tag{2.53}$$

We thus obtain the basic continuity equation for electrons as

$$\frac{\partial n}{\partial t} = \frac{1}{q}\cdot\frac{\partial J_n}{\partial t} + (G_n - R_n) \tag{2.54}$$

And similar continuity equation for holes as

$$\frac{\partial p}{\partial t} = \frac{1}{q} \cdot \frac{\partial J_p}{\partial t} + (G_p - R_p) \tag{2.55}$$

In addition to continuity equation, Poisson's equation

$$\frac{\partial E}{\partial x} = \frac{\rho}{\varepsilon} \tag{2.56}$$

must be satisfied, where ε is the semiconductor permittivity and ρ is the space charge density, which arises from the difference of charge neutrality equation.

2.10 BEHAVIOUR OF CHARGED PARTICLES AT THE INTERFACE

Interface is the boundary between two dissimilar materials such as solid and liquid, solid and vacuum, solid and solid, gas and solid etc. Out of them the interface between solid and solid, especially between a metal and a semiconductor is our point of interest because to get an electrical contact with a semiconductor a layer of metal has to be deposited over a semiconductor surface. In metal semiconductor interfaces, six different cases can arise depending upon the relative work function and the type of conductivity of the semiconductor.

2.10.1 Metal-n-type semiconductor with $W_m > W_s$

In this case the charge segregation would occur in such a way as to make the semiconductor positive and the metal negative as shown in Figure 2.11(a). The field set-up in this case is such that the electrons in the semiconductor face a retarding field and therefore the situation corresponds to a depletion layer in the semiconductor. From the Figure 2.11(a) it is clear that in order to move an electron from semiconductor to metal the amount of energy required is equal to the work function W_{sm} of the metal semiconductor contact. This effective work function is known as the barrier height and is eventually equal to the difference of the work function of the metal and the semiconductor plus the difference between the Fermi level of the semiconductor and the bottom of its conduction band. At equilibrium the flow of the carrier from metal to semiconductor or semiconductor to metal is zero. When a voltage (V) is applied the changes in the energy level is brought only on the semiconductor side and practically no change occurs in the metal side. Thus under forward bias when the metal is made positive, the barrier for the electrons in the semiconductor gets reduced by the applied potential as shown in Figure 2.11(b), thus allowing a large number of electrons to cross into the metal. Thus under forward bias a large current flows due to the electrons of the semiconductor towards the interface. On the otherhand at reverse bias the barrier for the electrons in the semiconductor increases as shown in Figure 2.11(c), so that only a small current flows across the interface. This type of interface thus, behaves as a rectifying interface, because it allows a large current to flow only in one direction.

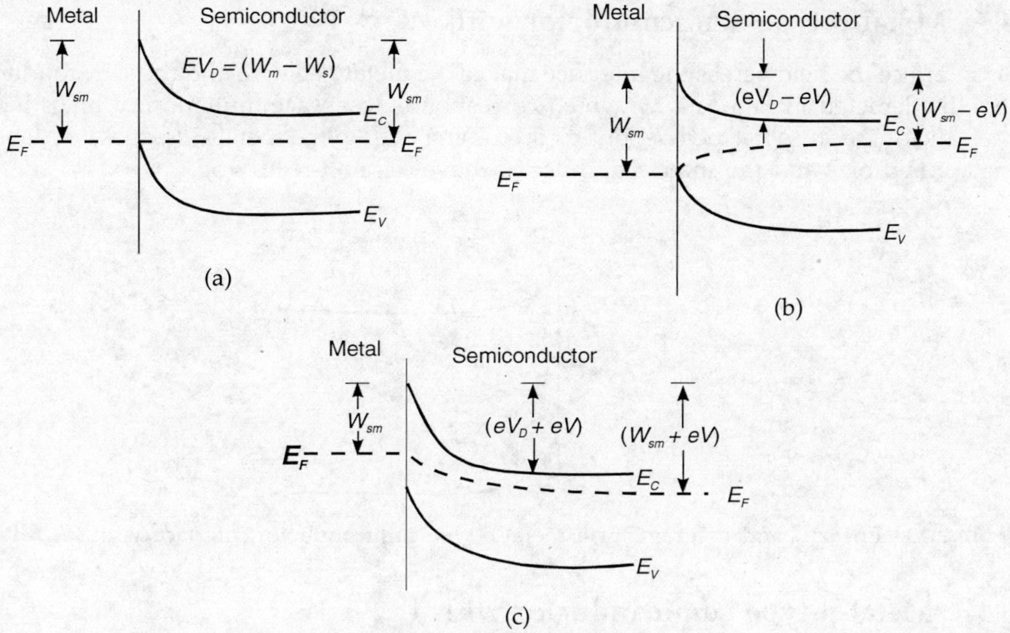

Figure 2.11 Energy band diagram of metal-n-type semiconductor (a) at equilibrium, (b) under forward bias, and (c) under reverse bias.

2.10.2 Metal-n-type semiconductor with $W_m < W_s$

Here exchange of the charge carrier makes the metal positive and the semiconductor negative as shown in Figure 2.12. From the field at the interface the majority carrier of n-type semiconductor will accelerate and thus it will create an accumulation layer in the semiconductor. Thus the crossing of these electrons from semiconductor to metal do not depend on the applied voltage (V) across the interface and the interface can be seen to be a non-rectifying one.

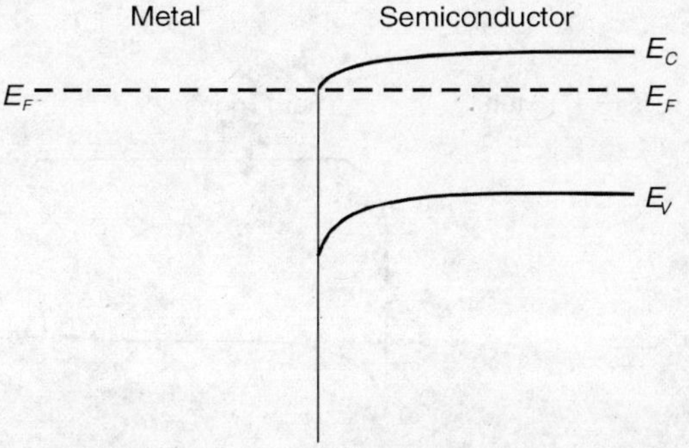

Figure 2.12 Energy band diagram for a metal-n-type semiconductor with $W_m < W_s$.

2.10.3 Metal-p-type semiconductor with $W_m > W_s$

In this interface the field across the interface makes the metal negative and the semiconductor positive as depicted in Figure 2.13, which corresponds to an accumulation of holes in the semiconductor (because it is p-type) but their crossing over to the metal side does not depend on the applied bias and the interface, thus it behaves as non-rectifying.

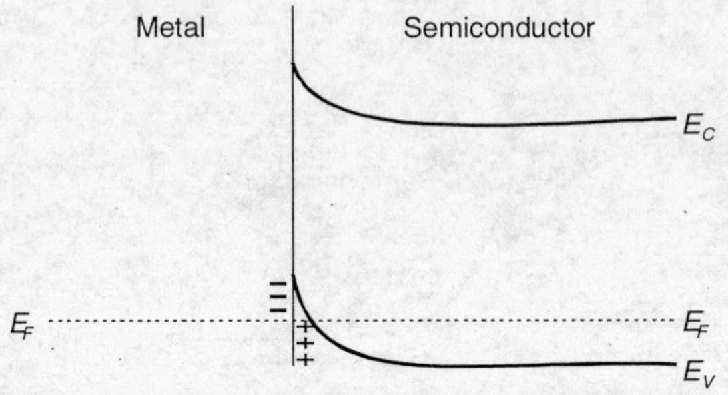

Figure 2.13 Energy band diagram for a metal-p-type semiconductor interface with $W_m > W_s$.

2.10.4 Metal-p-type semiconductor with $W_m < W_s$

Here the depletion of the hole takes place on the semiconductor side of the interface and thus the interface acts like a rectifying one. This case is similar to that of metal-n-type semiconductor with $W_m > W_s$. Energy band diagram of metal-p-type semiconductor interface at equilibrium, both at forward bias and at reverse bias are as shown in Figure 2.14(a), (b), and (c) respectively.

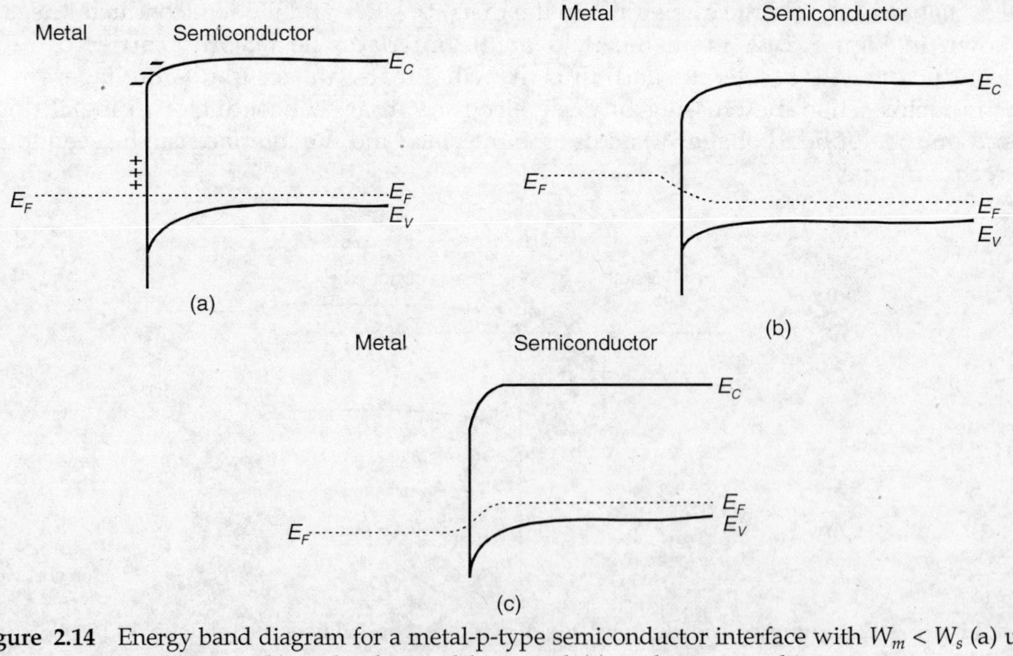

Figure 2.14 Energy band diagram for a metal-p-type semiconductor interface with $W_m < W_s$ (a) under equilibrium, (b) under forward bias, and (c) under reverse bias.

2.10.5 Metal intrinsic semiconductor interface

In this case the charge segregation takes place in such a way that it makes the metal negative and the semiconductor positive. As a result the field is set up in such a way that electrons of the semiconductor face a retarding field and holes face an aiding field, which results in the depletion of the electrons and the accumulation of holes on the semiconductor side. Under external bias, the net current across the interface is the sum of hole and electron currents of the two sides and is equal to zero. However, when the metal is made positive, the electron current from the semiconductor increases while the hole current does not change since they already face an aiding field. The electron current from the metal is very small and does not change with the applied bias. The interface therefore behaves as a poor rectifying one for $W_m > W_s$.

2.10.6 Potential distribution and space change width at metal semiconductor interface

Let for a metal-n-type semiconductor, N_D be the donor concentration and ε be the permittivity of the semiconductor and assuming that the depletion layer is an abrupt one, the Poisson's equation

$$\frac{d^2V(x)}{dx^2} = -\frac{eN_D}{\varepsilon} \tag{2.57}$$

or

$$d\left(\frac{dV(x)}{dx}\right) = -\frac{eN_D}{\varepsilon}\cdot dx$$

On integration

$$\frac{dV(x)}{dx} = \frac{-eN_D}{\varepsilon}x + c_1$$

Since at $x = 0$, field is maximum;

$$\therefore \quad \frac{dV(x)}{dx} = \left(E_{\max} - \frac{eN_D}{\varepsilon}x\right)$$

If W is the width of the depletion region, then the field

$$\frac{dV(x)}{dx} = \frac{eN_D}{\varepsilon}\cdot(W - x) \tag{2.58}$$

Integrating Equation (2.58) and using boundary conditions.

$$V(x) = -\varphi_{sm} \quad \text{at} \quad x = 0$$

and

$$V(x) = -\varphi_{sm} - (-V_D) = (-\varphi_{sm} + V_D) \qquad \text{at} \qquad x = W$$

where φ_{sm} is the voltage equivalent of the barrier height W_{sm}.

The potential at any point is given by

$$V(x) = \frac{eN_D}{\varepsilon}\left[Wx - x^2\right] - \varphi_{sm}$$

so that

$$V_D = \frac{eN_D}{2\varepsilon}W^2 \tag{2.59}$$

And therefore

$$W = \left(\frac{2\varepsilon}{eN_D}\cdot V_D\right)^{1/2} \tag{2.60}$$

The application of an external voltage V then changes the depletion layer width as

$$W = \left[\frac{2\varepsilon}{eN_D}(V_D - V)\right]^{1/2} \tag{2.61}$$

Polarity of V depends on the type of bias.

2.11 SEMICONDUCTOR JUNCTION

When two different semiconductor materials are contacted together a junction is formed. The semiconductor junctions are of two types—(a) homogeneous i.e., when a p-type silicon is connected to an n-type silicon, and (b) heterojunction, i.e., when a p-type silicon is combined with an n-type GaAs having different bandgap. In this Chapter we will discuss only about a single crystal semiconductor material containing both p-n junction. A p-n junction is extensively used in rectification, switching and in other operation in electronic circuits and is a key building block for the bipolar transistor and thyristor.

p-n junction can be produced by different methods and accordingly are known as (a) grown-junction, (b) alloyed junction, and (c) diffused junction.

(a) **Grown-junction:** When donor impurities are introduced into one side and acceptor impurities into the other side of a single crystal at the time of crystal growth from molten silicon or germanium, then a grown-junction is formed.

(b) **Alloy-junction:** Alloy type is made by melting a small pellet of indium on a slice of n-type germanium (or silicon). The molten indium is absorbed into the germanium (or silicon) and produces a p-type material. The area of contact between the pellet of indium and n-type semiconductor material is the alloy-junction.

(c) **Diffused-junction:** In this process a wafer of semiconductor material (say p-type) is exposed to a gas of pentavalent (n-type) impurity or a solid dopant source at a very high temperature inside a diffusion furnace. At this high temperature the energetic impurity atoms impinge on the p-type semiconductor wafer and forms the p-n junction. The depth of the junction depends on time and temperature of diffusion.

2.11.1 Unbiased p-n junction

A schematic diagram of a p-n junction is depicted in Figure 2.15. In n-type material the donor ion is positive, since it gives out an electron and itself becomes a positive ion. Similarly in p-type material, the acceptor is a negative ion. Thus in Figure 2.15 when the p-n junction is formed, p-side contains the holes and acceptor ions, on the other hand n-side contains electrons and the donor ions. Because of the density gradient across the junction, electron will initially diffuse across the junction to the left and holes to the right. In this process diffused holes and electrons will recombine and as a result the ions next to the junction will not have their accompanying mobile charges. Thus these ions will become uncovered charges. Figure 2.15(a) shows the situation after the initial diffusion takes place. The positive and negative uncovered charges generate an electric field called the barrier field, which opposes the diffusion of electrons and holes through the junction. At equilibrium the barrier field prevents further diffusion of electrons and holes, which causes no movement of charge carriers across the junction. Since the neighbourhood of the junction is depleted of mobile charges, it is referred to as depletion region (about 0.5 μm thick) or the space charge region or the transition region.

Figure 2.15(b) shows the variation of donors and acceptor ions around the junction. Figure 2.15(c) shows that the mobile charge concentration is zero in the depletion region. Again the uncovered charge density is zero at the junction; positive to its right and negative to its left in the depletion region [Figure 2.15(d)].

Poisson's equation gives the barrier field F as

$$\frac{d^2V}{dx^2} = -\frac{\rho}{\varepsilon} \tag{2.62}$$

where ε is the permittivity of the semiconductor and V is the electrostatic potential at a distance x and which can be obtained by integrating Equation (2.62)

$$F = -\frac{dV}{dx} = \int_{x_0}^{x} \frac{\rho}{\varepsilon} dx \tag{2.63}$$

The variation of barrier field across the junction is shown in Figure 2.15(e). The electrostatic potential at a point is given by

$$V(x) = -\int_{x_0}^{x} F \cdot dx \tag{2.64}$$

In Figure 2.15(e) the variation of $V(x)$ in the depletion region is shown. It rises in the n-side to give a potential barrier E_B that prevents further diffusion of carriers across the junction. Similarly in Figure 2.15(f) the potential barrier E_B prevents further diffusion of carriers right to leave.

Thus under unbiased condition there is no flow of holes or electrons across the junction and hence current flows through the junction when two electrodes of a bias battery are connected across the device, one to the p-side and the other to the n-side. Such a device is known as a p-n diode.

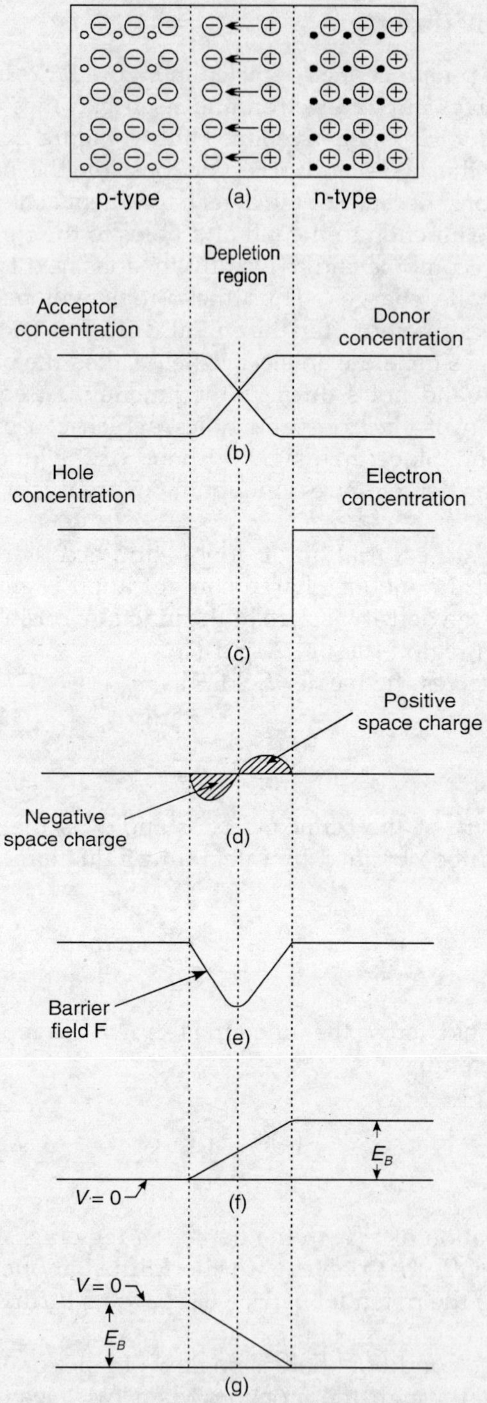

Figure 2.15 Schematic diagram of a p-n junction (a) mobile and immobile charges, (b) variation of donor and acceptor ion concentrations, (c) variation of electron and hole current, (d) concentration of uncovered charge, (e) electric field variation, (f) variation of electrostatic potential, and (g) electron energy variation.

2.11.2 Biased p-n junction

When the positive pole of a battery is connected to the p-side and the negative pole to the n-side of a p-n junction diode, then it is said to be at forward bias and when the above mentioned polarity are reversed then the diode is said to be at reverse bias.

(i) Forward biased p-n junction

A forward biased p-n junction and its circuit are depicted in Figure 2.16(a) and (b) respectively. Battery voltage *V* exerts a force on the holes of p-side and electrons of n-side. As a result they move towards the junction and reduces the width of the depletion layer and at the same time reduces the barrier height between the two conduction bands of p-n layers [Figure 2.18(a)]. This reduction in barrier height allows more holes to flow from p to n-side and more electrons from n to p-side, which results in more current flow through the junction. Here both the types of majority carriers of two sides, i.e., electron and hole flows are opposite to each other but the currents are in the same direction and constitute forward current. The forward bias does not affect the minority carrier current that flows in the direction opposite to the direction of the majority carrier current.

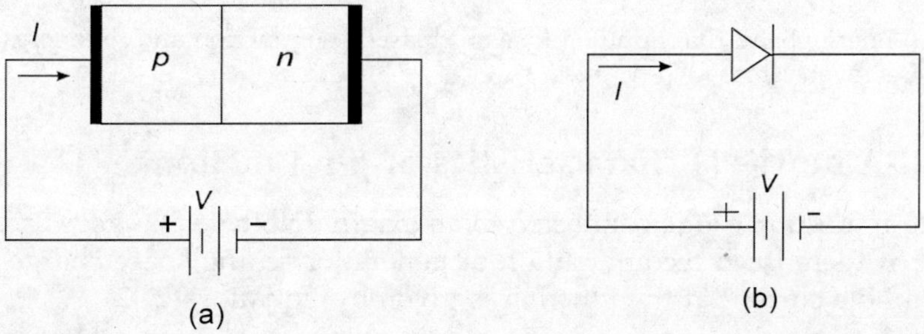

Figure 2.16 (a) Forward biased p-n junction and (b) circuit representation of forward biased p-n junction.

(ii) Reverse biased p-n junction

Figure 2.17(a) and (b) show respectively a reverse bias p-n junction and its circuit representation. The battery voltage *V* pulls the majority carrier holes and electrons on the two sides of p-n layers. As a result the holes and electrons will move away from the junction, which results in the increase of the depletion layer and rise in the barrier height. Thus due

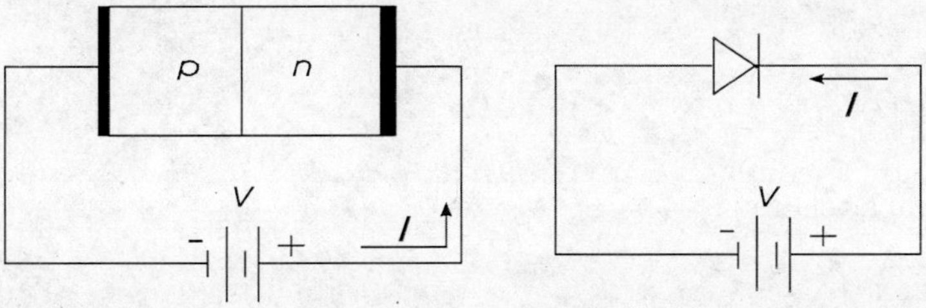

Figue 2.17 (a) Reverse biased p-n junction and (b) circuit representation of reverse biased p-n junction.

to the absence of the majority carriers the forward current becomes zero. Now at reverse bias the minority carriers will be repelled by the plates of the battery and as a result they move towards the junction. As here holes are minority carriers of n-layer and move towards the p-layer, thus the direction of current is now reversed.

As depletion region is devoid of free carriers, this region has a much higher resistivity than any part of the semiconductor. So most of the battery voltage V appears across the depletion region for both reverse and forward bias. The energy band diagram of forward biased and reverse biased junctions relative to the unbiased junction are shown in Figure 2.18(a) and (b) respectively.

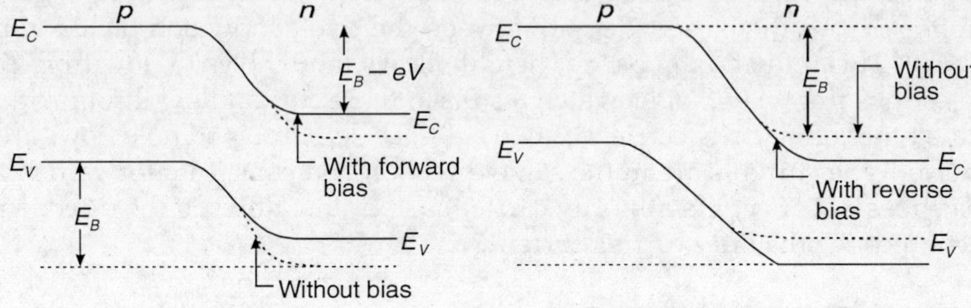

Figure 2.18 (a) Energy band diagram of a forward biased p-n junction and (b) energy band diagram of a reverse biased p-n junction.

2.11.3 Volt-Amp (V–I) characteristics of p-n junction

Let us take a p-n junction diode connected to an external voltage V at forward bias and also considering that there is no recombination or generation in the space charge region.

The total hole current at the junction is given by Equation (2.42)

$$J_p = qp\mu_p E - qD_p \frac{dp}{dx}$$

Both the terms on the right hand side of the above equation are much larger than the current J_p (the value of J_p is low for low level injection). Hence these two terms can be considered to be nearly equal, i.e.,

$$qp\mu_p E = qD_p \frac{dp}{dx}$$

or

$$E = \frac{D_p}{\mu_p p} \frac{dp}{dx} \tag{2.65}$$

Using the Einstein's relation $\quad \dfrac{D_p}{\mu_p} = V_T$

We have

$$E = \frac{V_T}{p} \frac{dp}{dx} \tag{2.66}$$

Now E is the electric field intensity and, therefore,

$$E = -\frac{dv}{dx} \tag{2.67}$$

Equating Equations (2.66) and (2.67), we get

$$\frac{V_T}{p}\frac{dp}{dx} = -\frac{dv}{dx}$$

or

$$\frac{dp}{p} = -\frac{dv}{V_T} \tag{2.68}$$

Let the concentration of holes under thermal equilibrium condition be p_{no} and that in the beginning of n-side is denoted by $p_n(0)$. Now integrating Equation (2.68) using the boundary condition, we get

$$p_n(0) = p_{no}\, e^{V/V_T} \tag{2.69}$$

Under low-level injection condition, the minority diffusion current (due to holes) crossing the junction of $x = 0$ is given by

$$I_{pn}(0) = \frac{AqD_p p'_n(0)}{L_p} \tag{2.70}$$

where

L_p = Diffusion length for the holes
A = Cross-section of the junction

and

$$p'_n(0) = p_n(0) - p_{no} \tag{2.71}$$

Substituting the expression for $p_n(0)$ from Equation (2.69) to Equation (2.71) we get

$$p'_n(0) = p_{no}\, e^{V/V_T} - p_{no} = p_{no}\,(e^{V/V_T} - 1) \tag{2.72}$$

Substituting Equation (2.72) in Equation (2.70) we get the expression of the current crossing the junction from p-side to n-side as

$$I_{pn}(0) = \frac{AqD_p}{L_p} p_{no}(e^{V/V_T} - 1) \tag{2.73}$$

The electron current $I_{np}(0)$ crossing the junction from n-side to p-side is obtained by interchanging the subscripts n and p, i.e.,

$$I_{np}(0) = \frac{AqD_n}{L_n} n_{po}(e^{V/V_T} - 1) \tag{2.74}$$

The total diode current is the sum of the hole current and electron current i.e.,

$$I = I_{pn}(0) + I_{np}(0) = \left[\frac{AqD_p p_{no}}{L_p} + \frac{AqD_n n_{po}}{L_n}\right]\left(e^{V/V_T} - 1\right)$$

or

$$I = I_0(e^{V/V_T} - 1) \tag{2.75}$$

where

$$I_0 = \left[\frac{AqD_p p_{no}}{L_p} + \frac{AqD_n n_{po}}{L_n}\right]$$

I_0 is known as the reverse saturation current. However for p-n junction diode when there is considerable amount of carrier generation and recombination, the current is given by

$$I = I_0(e^{V/\eta V_T} - 1) \tag{2.76}$$

$\eta = 1$ for germanium and $\eta = 2$ for silicon.

The Equation (2.76) can be expressed graphically and known as *V-I* characteristics of a diode.

(a) When applied voltage V is positive and $V >> V_T$, the term $e^{V/\eta V_T} >> 1$ and therefore Equation (2.76) becomes

$$I = I_0\, e^{V/\eta V_T} \tag{2.77}$$

So we get an exponential curve as shown in the first quadrant of the Figure 2.19.

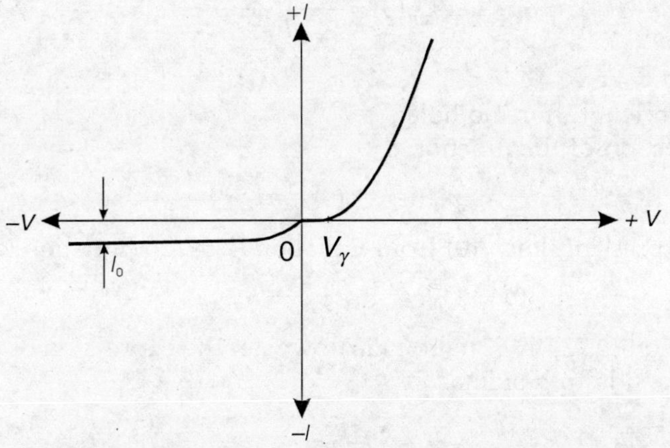

Figure 2.19 *V-I* characteristic of a p-n junction diode.

(b) When V is negative and $|V| >> V_T$, the term $e^{V/\eta V_T}$ becomes $e^{-V/\eta V_T}$ and, thus, $1/e^{V/\eta V_T} << 1$. Therefore Equation (2.76) becomes:

$$I = I_0\,(-1) = -I_0 \tag{2.78}$$

A straight line can represent this in the third quadrant of the Figure 2.19. This reverse saturation current is of minority carrier also known as leakage current.

(c) When $V = 0$, Equation (2.76) becomes

$$I = I_0\,(e^0 - 1) = I_0(1 - 1) = 0$$

So the curve passes through $V = 0$, $I = 0$, i.e., through the origin of the graph of Figure 2.19. V_γ is called the cut-in voltage of the diode at which the current starts to flow. For silicon p-n junction diode $V_\gamma = 0.7$ V and for germanium $V_\gamma = 0.3$ V. Cut-in voltage V_γ is defined as the minimum forward voltage at which 1% of the rated current flows.

As η for Ge and Si are not same, hence two sets of *V-I* characteristic curves are obtained as depicted in Figure 2.20. It has been formed that the reverse saturation current (I_0) increases approximately at the rate of 7% per degree centigrade for both silicon and germanium. Therefore, I_0 will increase 1.07 times, for every degree rise of temperature. For 10°C rise in temperature, the reverse saturation current will be $(1.07)^{10} \cong 2.0$ times more. In other words, I_0 will be double for every 10°C rise in temperature.

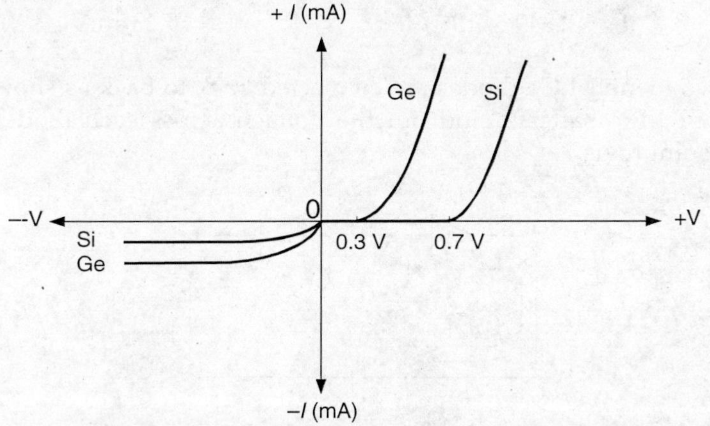

Figure 2.20 *V-I* characteristics of both silicon and germanium p-n junction diode.

If $I_0 = I_{01}$ at temperature T, thus at temp T_2

$$I_{02} = I_{01} \times 2^{(T_2 - T_1)/10} \tag{2.79}$$

If the temperature increases (with voltage remaining constant), the current increases. Now to maintain a constant current, voltage is to be reduced. The rate of voltage reduction is 2.5 mV/°C. In other words,

$$\frac{dV}{dT} = -2.5 \text{ mV/°C for constant } I.$$

Also V_T directly depends on temperature.

Problem 2.5 The voltage across a silicon diode at room temperature is 0.7 V when 2 mA current flows through it. If the voltage increases to 0.75 V, calculate the diode current. Asssume $V_T = 26$ mV.

Solution: From Equation (2.76) the forward current of a p-n junction

$$I = I_0 \left(e^{V/\eta V_T} - 1\right)$$

Given for silicon diode $\eta = 2$

$$V_T = 26 \text{ mV} = 0.026 \text{ V}$$

$$V_\gamma = 0.7 \text{ V}, I = 2 \text{ mA} = 2 \times 10^{-3} \text{ A}$$

Then

$$2 \times 10^{-3} = I_0(e^{0.7/2 \times 0.026} - 1)$$

or

$$I_0 = \frac{2 \times 10^{-3}}{e^{13.46} - 1} = 2.849 \times 10^{-9} \text{ A}$$

(b) When $V_\gamma = 0.75$ V

$$I = (2.849 \times 10^{-9})(e^{0.75/2 \times 0.026} - 1) = 5.23 \text{ mA}$$

Problem 2.6 Two identical Ge diodes are connected back to back as shown in Figure 2.21 with 10 V battery. Find out the voltage across each diode, making proper assumptions.

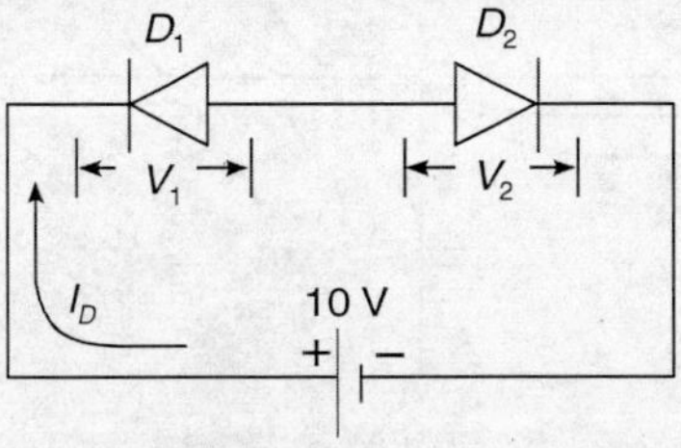

Figure 2.21 Two diodes are connected back to back.

Solution: As per the problem the required circuit arrangement is as shown in Figure 2.21. Here two diodes D_1 and D_2 are connected back to back and the total applied voltage 10 V will appear across the two diodes as V_1 and V_2. Thus because of the arrangements diode D_1 remains as reverse biased and D_2 as forward biased. As the two diodes are in series hence the current flowing through both of them must be same and equal to I_0, which is the reverse saturation current of the reverse biased diode D_1. Using Equation (2.76) current flowing through diode D_2 is

$$I = I_0[e^{V_2/\eta V_T} - 1]$$

Since for diodes D_2, $I = I_0$,

We have

$$I_0 = I_0[e^{V_2/\eta V_T} - 1]$$

As η for Ge is 1 then

$$I_0 = I_0 \, e^{V_2/V_T} - I_0$$

or

$$2I_0 = I_0 \, e^{(V_2/V_T)}$$

or

$$2 = e^{V_2/V_T}$$

Taking logarithm on both sides we get

$$\log_e 2 = \frac{V_2}{V_T}$$

or

$$V_2 = V_T \times 0.693$$

At room temperature V_T = 26 mV = 0.026 V

$$V_2 = 0.693 \times 0.026 = 0.018 \text{ V}$$

Thus the voltage V_1 across the diode D_1 is

$$V_1 = 10 - 0.018 = 9.982 \text{ V}$$

2.11.4 Junction capacitances

In a semiconductor p-n junction, two types of capacitance comes into the picture:

(a) **Transition capacitance:** This capacitance arises in a reverse biased p-n junction and also known as space charge capacitance.
(b) **Diffusion capacitance:** This has negligible value at reverse biased condition but has a large value at forward biased condition.

Transition capacitance (C_T)

When a p-n junction is reverse biased majority carriers move away from the junction, thus uncovering more number of immobile charges. Hence thickness of the depletion layer increases with reverse bias voltage. This increase in uncovered charge with applied voltage is equivalent to a capacitance and is called a transition capacitance.

Thus

$$C_T = \frac{dQ}{dV}$$

and current

$$i = \frac{dQ}{dt} = \frac{dQ}{dV}\frac{dV}{dt} = C_T \frac{dV}{dt}$$

When the p-n junction is an alloy junction it becomes an abrupt type. In any p-n junction, in practice the concentration of acceptor ions (N_A) is much larger than the concentration of donor ions (N_D), i.e., $N_A >> N_D$.

When a p-n junction is reverse biased, the uncovered charges will appear on both sides of the junction but net charge must be zero, then

$$q\, N_A\, W_p = q\, N_D\, W_n \tag{2.80}$$

where

W_p = Width of the depletion region in p-side
W_n = Width of the depletion region in n-side
Q = Charge of the carrier.

Total width of the depletion layer $W = W_p + W_n \cong W_p$

Now from Poisson's equation

$$\frac{d^2V}{dx^2} = \frac{-\rho}{\varepsilon} = \frac{-qN_A}{\varepsilon} \tag{2.81}$$

Integrating the above equation twice with the initial conditions:

at $x = W_p \cong W$; $V = V_j$ we get

$$V_j = \frac{qN_A W^2}{2\varepsilon} \tag{2.82}$$

Therefore

$$W^2 = \frac{2\varepsilon V_j}{qN_A} \tag{2.83}$$

The reverse bias voltage V across the p-n junction increases the barrier potential to V_0, thus the junction voltage

$$V_j = V + V_0 \tag{2.84}$$

Substituting Equation (2.84) in Equation (2.83), we get

$$W^2 = \frac{(V_0 + V)2\varepsilon}{qN_A} \tag{2.85}$$

Differentiating Equation (2.85) w.r.t. V, we obtain

$$2W\frac{dW}{dV} = \frac{2\varepsilon}{qN_A}$$

or

$$\frac{dW}{dV} = \frac{\varepsilon}{WqN_A} \tag{2.86}$$

If the cross-sectional area of the junction be A, then quantity of charge crossing the area A

$$Q = qN_A\, W\, A \tag{2.87}$$

Now from the definition of transition capacitance

$$C_T = \frac{dQ}{dV} = \frac{d}{dV}(qN_A AW) \qquad \text{(from Equation 2.87)}$$

$$= qN_A A\frac{dW}{dV} \tag{2.88}$$

Substituting Equation (2.86) in Equation (2.88), we get

$$C_T = qN_A A\frac{\varepsilon}{W\,qN_A}$$

or

$$C_T = \frac{\varepsilon A}{W} \tag{2.89}$$

Diffusion capacitance (C_D)

During forward biased condition, the diffusion capacitance (C_D) comes into existence. If the change in the forward biased voltage of this diode dV and the corresponding change in charge due to injected carrier is dq, then the diffusion capacitance

$$C_D = \frac{dq}{dV}$$

Typical value of diffusion capacitance in the forward direction ranges from several nanofarads to several microfarads. Obviously it is many times larger than the transition capacitance and thus at forward bias C_D swamps out any effect due to C_T.

2.11.5 Effects of high field on charge carriers in semiconductors

The drift velocity of charge carriers follows a linear relationship with electric field according to Ohm's law only for low values of electric field up to 10^5 V/m. The measured value of drift velocity of electrons and holes against electric field is shown in Figure 3.3 of Chapter 3 on IMPATT diodes. It is seen that the linear rise of the drift velocity with electric field changes at higher field and at a very high field $E > 10^6$ V/m the drift velocity saturates and becomes almost constant. The saturation of the drift velocity is an important phenomena underlying the action of many microwave semiconductor devices like IMPATT diodes, TRAPATT diodes, Gunn diodes, etc. The saturation of drift velocity has been explained in terms of optical phonon scattering. When the two faced centred cubic sub-lattices of a semiconductor crystal (e.g., silicon) undergo thermal vibrations in opposite direction, the quantum of vibration is called optical phonon, which has much higher energy (0.063 eV for Si, 0.37 eV for Ge, 0.35 eV for GaAs) than the acoustical phonon which is the quantum of the vibration when the sub-lattices move in the same direction. At low electric field, the carriers gain energy from electric field and loose energy to the vibrating lattice by emitting acoustic phonons and the velocity in the case increases linearly with electric field. But at high electric field, the electrons get sufficient energy between successive collisions to loose energy to the lattice by emitting optical phonons. The energy transfer to the lattice at high electric field is thus efficiently carried out by optical scattering, which leads to the saturation of the drift velocity, which will be explained now.

Let the relaxation time for optical phonon scattering of carriers in a semiconductor subjected to a high field be τ_0. The mobility in the high field case can be written as

$$\mu_0 = \frac{q\tau_0}{m^*} \tag{2.90}$$

where m^* is the effective mass of the carrier and q is the electronic charge.

The carrier looses an energy E_0 in each optical phonon collision by emitting an optical phonon. Thus the rate of energy loss of the carrier by optical phonon scattering is given by

$$\frac{dE}{dt} = \frac{E_0}{\tau_0} \tag{2.91}$$

The rate of energy gain of the carrier from the electric field is

$$\frac{dE}{dt} = q(\mu\varepsilon)\varepsilon \tag{2.92}$$

where $\mu\varepsilon$ denotes the path travelled in the direction of the field per unit time and ε denotes the value of the electric field.

The rate of energy gain is balanced by the rate of energy loss at equilibrium. Thus

$$q\mu\varepsilon^2 = \frac{E_0}{\tau_0} \tag{2.93}$$

or

$$q\frac{q\tau_0}{m^*}\varepsilon^2 = \frac{E_0}{\tau_0}$$

or

$$\tau_0 = \frac{(E_0 m^*)^{1/2}}{q\varepsilon} \tag{2.94}$$

The mobility μ in the high field case is thus inversely proportional to the electric field ε

$$\mu_0 = \frac{q\tau_0}{m^*} = \left(\frac{E_0}{m^*}\right)^{1/2} \frac{1}{\varepsilon} \tag{2.95}$$

Consequently the drift velocity $v_d = \mu\varepsilon$ becomes constant and independent of the field in the high field case. It is called saturated drift velocity v_s and is given by

$$v_s = \left(\frac{E_0}{m^*}\right)^{1/2} \tag{2.96}$$

A more accurate analysis shows that the drift velocities for Ge and Si approach a saturation velocity

$$v_s = \sqrt{\frac{8E_0}{3\pi m_0}} \tag{2.97}$$

For GaAs, the velocity field characteristic shows a region of negative differential mobility between 3×10^3 V cm^{-1} and 5×10^4 V cm^{-1} before it reaches saturation at high field. This is caused by inter-valley scattering of electrons between two valleys in the GaAs conduction band having different values of effective masses for electrons. These phenomena will be further discussed while discussing transfer electron devices.

2.11.6 Impact ionization and creation of e-h pairs at high field

At high field the electrons may gain enough energy from the field and reach ionization threshold energy E_i to make an ionization collision thereby lifting an electron from the valence band to the conduction band and producing an electron hole pair. The ratio of mean free path for ionizing collision l_i to the mean face pattern for optical collision l_o at ionization threshold energy is typically 20 to 50. Ionization rate is an important parameter for understanding the impact ionization. Ionization rate α_n/α_p of electron/hole is defined as the number of electron-hole pairs generated by an electron/hole per unit distance of travel in the direction of the field.

Ionization rates α_n, α_p increase very strongly with electric field. The measured results for Ge, Si and GaAs can be fitted to the Equation (2.98).

$$\alpha_{n,p} = A_{n,p} \exp\,(-b_{n,p}/\varepsilon)^m \tag{2.98}$$

where

$A_n = 1.55 \times 10^7$ cm^{-1} for Ge, 3.8×10^6 cm^{-1} for Si and 3.5×10^5 cm^{-1} for GaAs
$A_p = 1.0 \times 10^6$ cm^{-1} for Ge, 2.25×10^7 cm^{-1} for Si and 3.5×10^5 cm^{-1} for GaAs

and

$b_n = 1.56 \times 10^6$ V cm^{-1} for Ge, 1.75×10^6 V cm^{-1} for Si and 6.85×10^5 V cm^{-1} for GaAs
$b_p = 1.98 \times 10^6$ V cm^{-1} for Ge, 3.27×10^7 V cm^{-1} for Si and 6.85×10^5 V cm^{-1} for GaAs

2.11.7 Theory of exponential field dependence of ionization rate

Shockley [2] considered a simple model to explain field dependence of ionization rate. He assumed that only that electron produces ionization collision, which gains energy up to E_i without making a single optical phonon collision. To gain energy ionization, threshold energy E_i from the electric field ε, an electron should travel a path length of $l = E_i/q\varepsilon$. Since $E_i >> E_o$, the electron has energy to emit an optical phonon for most of this path length. Therefore the probability that this electron reaches E_i without suffering optical phonon collision is

$$P_{ro} = \exp\left(-\frac{l}{l_o}\right) \tag{2.99}$$

After reaching the ionization threshold energy the probability that the electron makes an impact ionization avoiding a likely optical phonon collision is given by

$$P_r = \frac{l_o}{l_i}\exp\left(-\frac{l}{l_o}\right) \tag{2.100}$$

Thus the electron must make $1/P_r$ starts from zero energy to make one ionizing collision. The other starts $(1/P_r - 1)$ produce an optical phonon collision when the electron drops back to zero energy. Thus the field supplies a total energy of $E_T = E_i + E_o\,(1/P_r - 1)$ to produce a single ionizing collision. Hence we have

$$E_T = E_i + E_o\,\{l_i/l_o \exp\,(l/l_o) - 1\}$$

Now per unit distance α, e-h pairs are created where α is the ionization rate supplied per ionization collision. It can also be written as

$$E_T = \frac{q\varepsilon}{\alpha}$$

Thus

$$\alpha = q\varepsilon\left[\frac{1}{E_i + E_o\left[\frac{l_i}{l_o}\exp\frac{l}{l_o} - 1\right]}\right]$$

where $E_o\left[\frac{l_i}{l_o}\exp\frac{l}{l_o} - 1\right]$ is much greater than E_i

and

$$\frac{l}{l_o} = \frac{q\varepsilon l}{q\varepsilon l_o} = \frac{E_i}{q\varepsilon l_o}$$

Thus we can write an approximate expression for α

$$\alpha = \left(\frac{q\varepsilon l_o}{E_o l_i}\right)\exp\left(-\frac{E_i}{q l_o \varepsilon}\right) \tag{2.101}$$

It shows that α has an exponential dependence on ε and varies very strongly with increase in the electric field.

2.11.8 Charge multiplication and avalanche breakdown of junctions

The depletion layer on either side of the metallurgical junction of a p-n diode is subjected to high field when it is reverse biased. With increasing reverse bias the electric field increases and becomes large enough to cause very large charge multiplication by impact ionization and the small saturation current sharply increases which can be observed in the *V-I* characteristics of a p-n junction diode. These phenomena can be understood in terms of ionization rate of charge carriers at high electric field and the condition of avalanche breakdown can be derived. Consider a reverse biased p-n junction with depletion layer around the junction plane of width W as shown in Figure 2.22, which is subjected to a high dc reverse field. A small reverse hole current I_{po} increases from left edge of the depletion layer by avalanche multiplication and reaches a value $M_p\, I_{po}$ at the right edge. Similarly dc reverse electron current increases from the right edge of the depletion layer from I_{no} and becomes $M_n I_{no}$ at the left edge due to multiplication. M_p and M_n are the current multiplication factors for holes and electrons respectively. The total dc current I_o is constant at the steady state being the sum of hole current and electron current $I_p + I_n = I_o$. The small increase of hole at x within an increment Δx inside the depletion layer is equal to the number of electron hole pairs generated per second in Δx.

Thus

$$\Delta\left(\frac{I_p}{q}\right) = \left(\frac{I_p}{q}\right)(\alpha_p\,\Delta x) + \left(\frac{I_n}{q}\right)(\alpha_n\,\Delta x)$$

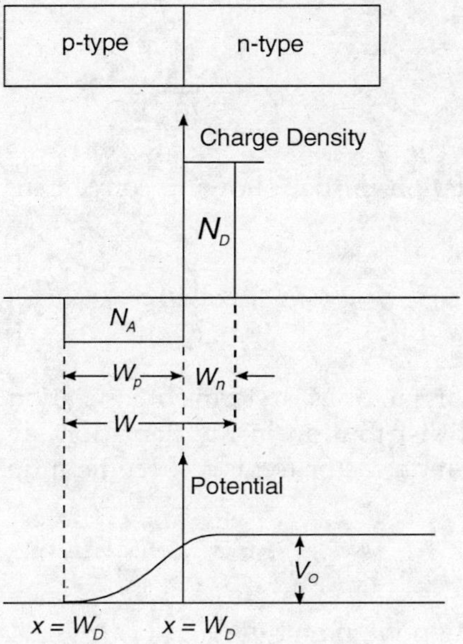

Figure 2.22 The charge density and potential variation at a p-n junction.

or

$$\frac{\Delta I_p}{\Delta x} - (\alpha_p - \alpha_n)\, I_p = \alpha_n I$$

The standard solution of the equation is given by

$$I_p = \frac{\int_0^W \alpha_n I \left[e^{\int_0^x -(\alpha_p - \alpha_n)\,dx} \, dx \right] + C}{e^{\int_0^x -(\alpha_p - \alpha_n)\,dx}}$$

where C is the constant of integration

Now the boundary conditions are

$$I_p = I_{po} \text{ at } x = 0$$

and

$$I_p = M_p I_{po} \text{ at } x = W$$

Thus the solution of the equation becomes

$$I_p(x) = \frac{I_0 \left\{ \frac{1}{M_p} + \int_0^x \alpha_n \exp\left[-\int_0^x (\alpha_p - \alpha_n)\,dx \right] dx \right\}}{\exp\left[-\int_0^x (\alpha_p - \alpha_n)\,dx \right]}$$

since

$$M_p = \frac{I_p(W)}{I_p(0)}$$

One can write the above equation in the following convenient form

$$1 - \frac{1}{M_p} = \int_0^W \alpha_p \exp\left[-\int_0^x (\alpha_p - \alpha_n)\, dx\right] dx \tag{2.102}$$

which relates the multiplication factor M_p in terms of ionization rates α_p and α_n which in turn is a function of electric field distribution in the depletion layer. At avalanche breakdown voltage the current multiplication factor tends to become infinite. Thus

$$M_p \to \infty \qquad \text{and} \qquad \frac{1}{M_p} \to 0$$

Hence the breakdown condition is given by

$$\int_0^W \alpha_p \exp\left[-\int_x^W (\alpha_p - \alpha_n)\, dx\right] dx = 1$$

This integral on L.H.S. is called the ionization integral.

If the hole initiates the ionization process instead of electrons the ionization integral is given by

$$\int_0^W \alpha_p \exp\left[-\int_x^W (\alpha_p - \alpha_n)\, dx\right] dx = 1$$

Either of these integrals give the breakdown condition since a mixed primary current initiate the ionization process.

If the two ionization rates α_n and α_p are equal in a semiconductor such as GaP the breakdown condition becomes

$$\int_0^W \alpha\, dx = 1 \tag{2.103}$$

when $\alpha = \alpha_n = \alpha_p$

The field distribution of a reverse biased p-n junction can be obtained from its doping profile. Using the data for field variation of ionization rates and the above-mentioned breakdown condition one can obtain the average breakdown voltage by numerically integrating the field over the depletion layer. The breakdown voltage for one-sided abrupt junction of Ge, Si and GaAs and GaP has been commuted against impurity concentration by Sze and Gibbons [3] and is as shown in Figure 2.23. The corresponding depletion layer width and maximum field at breakdown is shown in Figure 2.24.

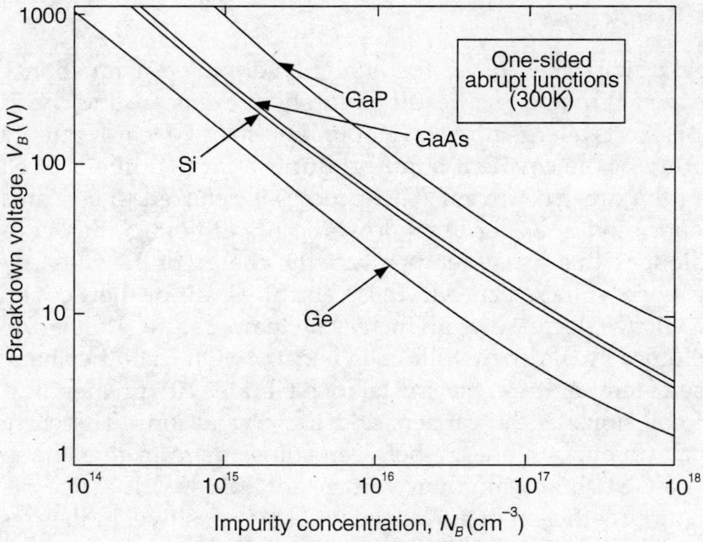

Figure 2.23 Avalanche breakdown voltage versus impurity concentration for one sided abrupt junctions in Ge, Si, GaAs and GaP [3].

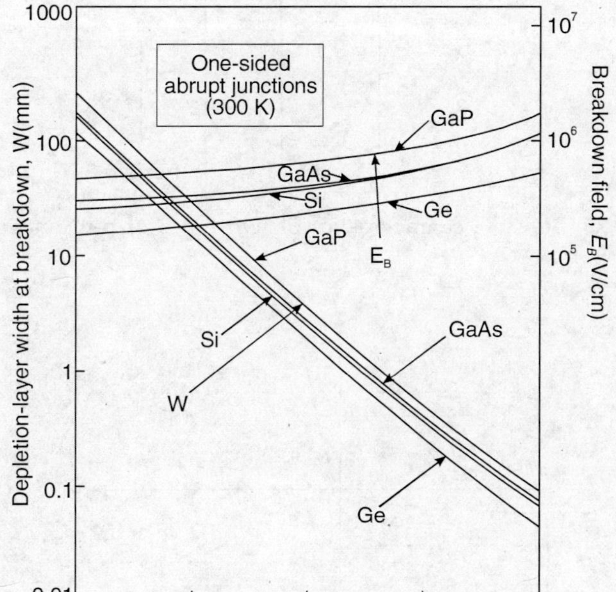

Figure 2.24 Depletion layer width and maximum field at breakdown for one-sided abrupt junction in Ge, Si, GaAs and GaP [3].

2.11.9 Zener Breakdown

This type of breakdown takes place for a highly doped p-n junction diode, where the depletion layer is very thin and as a result a high field exists around the junction. This high field at the junction exerts a large force on a bound electron to tear it out of its covalent bond. Thus a direct rapture of the covalent bond produces a large number of electron hole pairs, thereby increasing the reverse current. This process is referred to as Zener breakdown and the diodes are Zener diodes. Zener breakdown occurs at voltage below 6 V.

In a Zener diode with a rise of temperature the energy of the valence electron increases so that a lower applied voltage can pull these electrons out of their covalent bonds. So the Zener breakdown voltage decreases with increasing temperature. That is why the temperature coefficient of the Zener breakdown voltage is negative. On the other hand in an avalanche diode with the rise of temperature, the crystal ions vibrate with greater amplitude, increasing the possibility of collisions of the carriers and the crystal ions. The carrier thus have less opportunity to pick up enough energy between collisions to trigger the avalanche process. Consequently the avalanche breakdown voltage increases with a rise of temperature and hence the temperature coefficient of avalanche voltage is positive. In Figures 2.25 and 2.26 the reverse characteristics of avalanche diode and Zener diode are as shown.

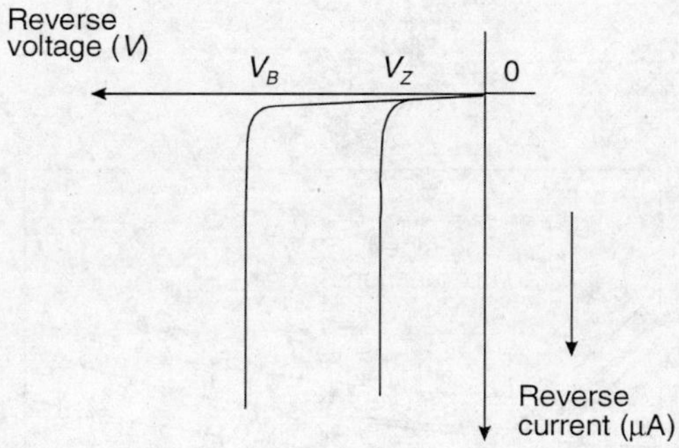

Figure 2.25 Reverse characteristics of two types of p-n junction diode.

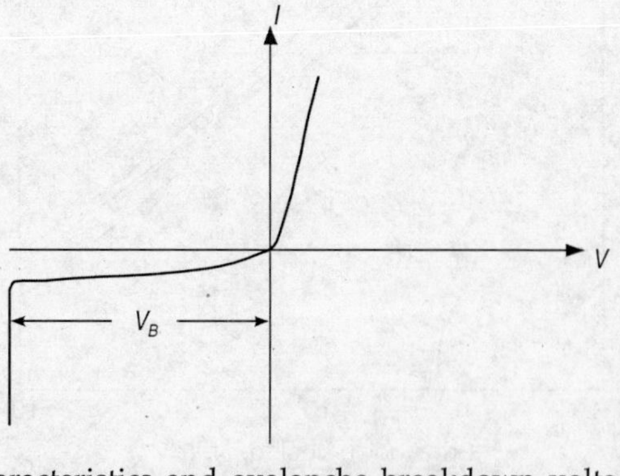

Figure 2.26 *V-I* characteristics and avalanche breakdown voltage of a p-n junction.

2.11.10 Resistance of a p-n junction diode

As Ohm's law is not valid for a p-n junction diode, hence its voltage current characteristics are non-linear. At any point on this graph the ratio of voltage and current gives the static or dc resistance of the diode

i.e.,
$$r_{dc} = \frac{V}{I} \tag{2.104}$$

As the graph is non-linear, thus r_{dc} varies from point to point of the graph.

The inverse of the slope of the volt-ampere characteristics of the p-n diode defines the dynamic or the ac resistance of the diode, thus

$$r_{ac} = \frac{dV}{dI}$$

which is not a constant and is determined by the operating voltage.

REFERENCES

1. Smith, R.A., *Semiconductors*, 2nd ed., Cambridge University Press, London, 1979.
2. Shockley, W., Problems related to p-n junction in Silicon, *Solid State Electronics*, Vol. 2, p. 35, 1961.
3. Sze, S.M. and Gibbons, G., Avalanche breakdown voltage of abrupt and linearly graded p-n junction in Ge, Si, GaAs and GaP, *Applied Physics Letter*; Vol. 8, p. 111, 1966.

SUGGESTED READINGS

1. Sze, S. M., *Semiconductor Devices*, 2nd ed., John Wiley & Sons, 2002.
2. Millman, J. and Halkias, C. C., *Electronic Devices and Circuits*, 2nd ed., Tata McGraw-Hill, New Delhi, 1996.
3. Smith, R.A., *Semiconductors*, 2nd ed., Academic Publishers, Calcutta, 1989.
4. Chattopadhyay, D. and Rakshit, P.C., *Electronics Fundamentals and Applications*, 3rd ed., New Age International, New Delhi, 2001.

PROBLEMS

1. With the help of energy band diagram explain the differences between a metal, a semiconductor and an insulator.
2. Write the Fermi-Dirac distribution function and explain its importance. What is Fermi level?
3. An ion has a positive charge numerically equal to twice the electric charge. The mass of the ion is 7360 times that of an electron. The ion is initially stationary and accelerated through a potential difference of 2 kV. Calculate the velocity and the kinetic energy acquired by the ion.
4. Write down the differences between an intrinsic and extrinsic semiconductor. How can an intrinsic semiconductor be an extrinsic one?

5. How an n-type semiconductor differs from a p-type one?
6. When can a semiconductor be degenerate?
7. A donor impurity is added to intrinsic silicon and the resistivity at room temperature is found to be 9.6 ohm·cm. Calculate the ratio of donor atoms to silicon atoms per unit volume.
8. What is the position of Fermi level in an intrinsic semiconductor? How does its position change when (a) donors and (b) acceptors are added to the semiconductor? What happens if the temperature is raised?
9. What do you mean by the lifetime of minority carriers in an impurity semiconductor? Why does the lifetime decrease when traps are introduced into the forbidden gap?
10. At room temperature Ge has an intrinsic concentration of 2.5×10^9 m^{-3}. If the electron mobility is 0.18 $m^2/(V \cdot s)$, calculate the electrical conductivity of intrinsic Ge. Also calculate the diffusion constants for electrons and holes.
11. The intrinsic carrier density of Ge at room temperature is 2.0×10^{19} m^{-3}. It is doped with As atom with a concentration of one As atom per million atoms of Ge. If the density of Ge atom is 4×10^{28} m^{-3}, determine the conductivity of the semiconductor. Assume complete ionization of As atoms and take $\mu_m = 0.35$ $m^2/(V \cdot s)$.
12. Explain the phenomenon of diffusion of current carriers in a semiconductor and write down the Fick's Law in connection to this.
13. Write down the conductivity equation and explain its significance.
14. Discuss about different types of metal semiconductor devices.
15. Define the following terms:

    (a) built-in potential, (b) barrier potential, and (c) bandgap of a semiconductor.
16. What is a p-n junction diode? How does a barrier field appear across a p-n junction?
17. What is an abrupt and a linearly graded p-n junction?
18. With the help of energy band diagram explain the operation of a p-n junction.
19. What is the origin of the reverse saturation current in a p-n junction? How reverse saturation current varies with temperature and biasing voltage?
20. Define the ac and dc resistance of a p-n junction diode. Are they equal? If not, why?
21. Write down the difference between avalanche breakdown and Zener breakdown.
22. Two identical silicon diodes are connected in series with 10 V dc source having 100 Ω internal resistance. Determine the current and voltage across each diode if the diodes are connected

    (i) back to back
    (ii) both diodes forward biased
    (iii) both diodes reverse biased.

    Assume suitable data.

23. The leakage current in a p-n junction diode is 10 μA at 20°C. Calculate the change in temperature required so that the leakage current may rise to 35 μA.

24. The reverse current of Si diode is found to be 50 nA at 27°C with 10 V reverse voltage. In a certain circuit the junction temperature was observed to rise up to 105°C. Calculate the maximum value of reverse current.

25. Explain why Zener breakdown voltage has a negative temperature coefficient, whereas the avalanche breakdown voltage has a positive temperature coefficient.

26. The reverse saturation current at room temperature of a p-n junction Ge diode is 10 μA. Find the voltage to be applied across the junction to obtain a forward current of 60 mA.

27. Calculate the ratio of the current for a forward bias of 0.08 V to the current for the same value of reverse bias applied to a Ge p-n diode at 27°C.

Chapter 3

Microwave IMPATT Devices

3.0 INTRODUCTION

IMPATT diodes have emerged, at the end of the twentieth century, as very powerful solid-state sources of microwaves, millimeter waves, and sub-millimeter waves, covering a wide range of frequency spectrum. The reported frequency range of IMPATT oscillators extend from below X-band (6 GHz) to the sub-millimeter wave region. The CW power output from a single device is quite high and reported results indicate 12 W at 6 GHz [1], 1 W at 94 GHz, and 2.2 mW at 412 GHz [2]. The vast frequency range and high power output should make the IMPATT a highly suitable device to meet the ever-increasing communication needs of the world. IMPATTs have been used in microwave and millimeter wave (digital and analog) communication system, in radars for civilian purposes, and in missiles for defence systems.

IMPATT is the acronym for IMPact Avalanche Transit Time, which reflects the mechanism of its operation. In its simplest form an IMPATT is a p-n junction diode reverse-biased to breakdown, in which an avalanche of electron-hole pairs is produced in the high field region of the device depletion layer by impact ionization. The transit of the carriers through the depletion layer leads to the generation of microwaves and millimeter waves when the device is tuned in a microwave or a millimeter wave cavity as shown in Figure 3.1. Microwave oscillations from a simple p^+nn^+ silicon diode reverse-biased to breakdown were first reported in 1965 by Johnston, Deloach and Cohen [3] from Bell Laboratories, but the same was earlier predicted by Read [4] for a special doping profile (p^+nn^+) as shown in Figure 3.2. It was later found that p-n junctions of various doping profiles based on any semiconductor would give rise to IMPATT oscillations. So far IMPATT oscillations have been observed from Ge, Si, GaAs and InP diodes. The thin depletion layer of an IMPATT diode is subjected to high field at breakdown voltage and the important high field phenomena underlying IMPATT action are: (1) saturation of drift velocity of charge carriers and (2) charge multiplication by impact ionization.

3.1 SATURATION OF DRIFT VELOCITY

In most semiconductors the drift velocity increases linearly with electric field according to Ohm's law in the low field range, when the acoustic phonons are the main scattering agencies. At high field ($E > 10^6$ V/m) the dominant scattering mechanism is due to optical

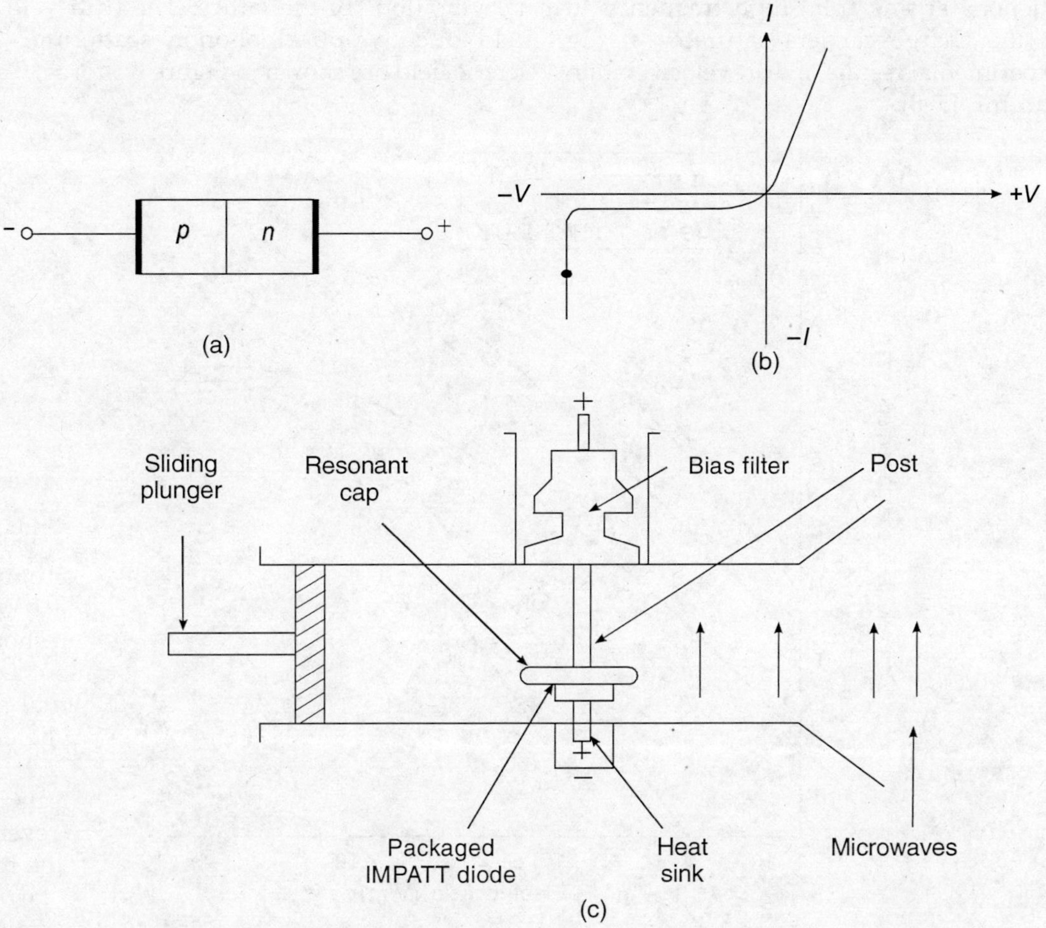

Figure 3.1 Schematic diagram of (a) basic IMPATT structure, (b) dc operating point, and (c) microwave IMPATT oscillator.

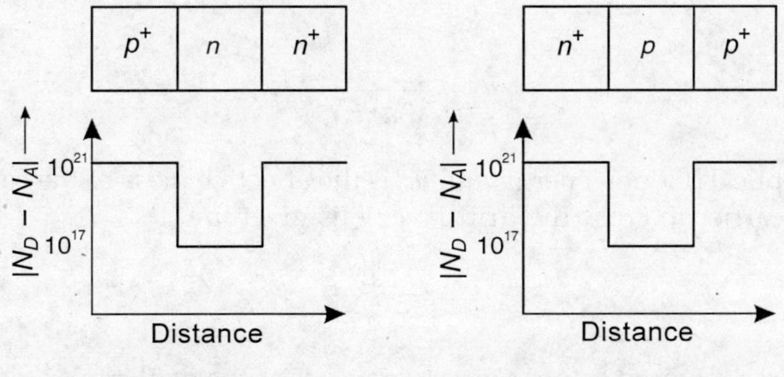

Figure 3.2 Structure of SDR IMPATT diode.

phonons arising from high frequency thermal vibrations of the lattice. The drift velocity of the charge carriers saturates at high field due to optical phonon scattering. The experimental results of drift velocity against electric field are shown in Figure 3.3 for Si, GaAs and InP [5–8].

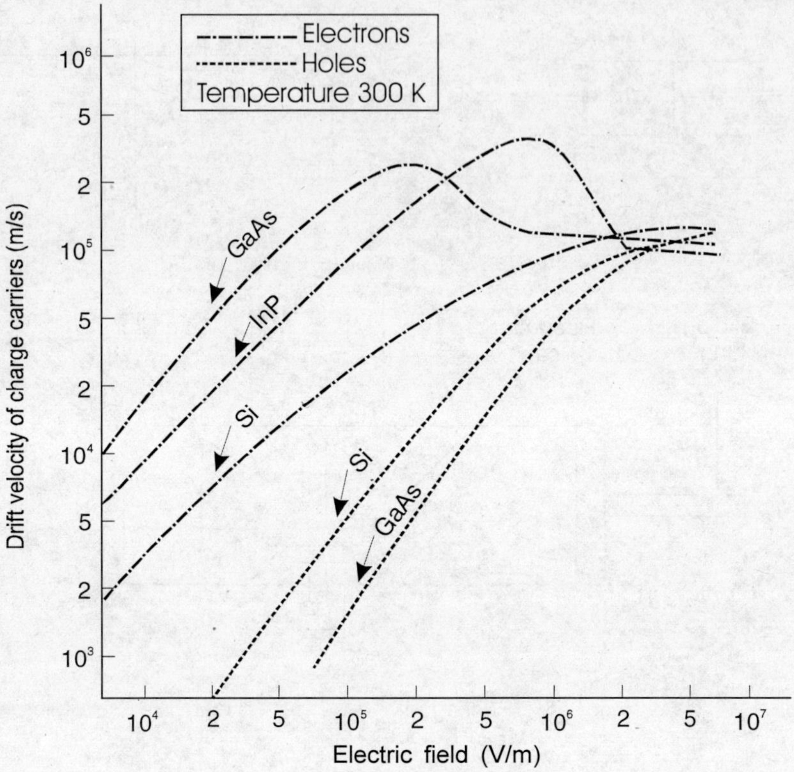

Figure 3.3 Drift velocity versus electric field for electron in Si, and GaAs and InP; and for holes in Si and GaAs

An approximate expression for the saturated drift velocity of charge carriers, v_s is given by

$$v_s = \frac{8E_{op}}{(3\pi m^*)^{1/2}} \tag{3.1}$$

where E_{op} is the optical phonon energy and m^* is the effective mass of the carriers. The transit time of a charge carrier to cross the drift layer L is given by

$$\tau = \frac{L}{v_s} \tag{3.2}$$

Thus the transit time required by charge carriers to cross the depletion region is proportional to the width of the depletion layer, since the velocity is constant at high field.

3.2 IMPACT IONIZATION, CARRIER MULTIPLICATION AND JUNCTION BREAKDOWN

At higher fields (10^7 V/m and above) a carrier may gain enough energy from the electric field to reach the ionization threshold, when it can make an impact on an electron bound in the valence band and lift it to the conduction band, thereby creating a mobile electron in the conduction band and a mobile hole in the valence band. This process is known as the creation of an electron hole (e-h) pair by impact ionization. The ionization rates α_n and α_p for electrons and holes are defined to be the number of e-h pairs produced by an electron or hole per unit distance of travel in the direction of the electric field and are very strong functions of the electric field. The experimental results are approximately fitted with the empirical formula

$$\alpha_{n.p} = A_{n.p} \exp \frac{-B_{n\cdot p}}{E} \tag{3.3}$$

where $A_{n.p}$ and $B_{n.p}$ are constants for a semiconductor. Ionization rates for IMPATT related semiconductor (Ge, Si, GaAs and InP) have been measured by several workers and are shown in Figure 3.4. [9-11].

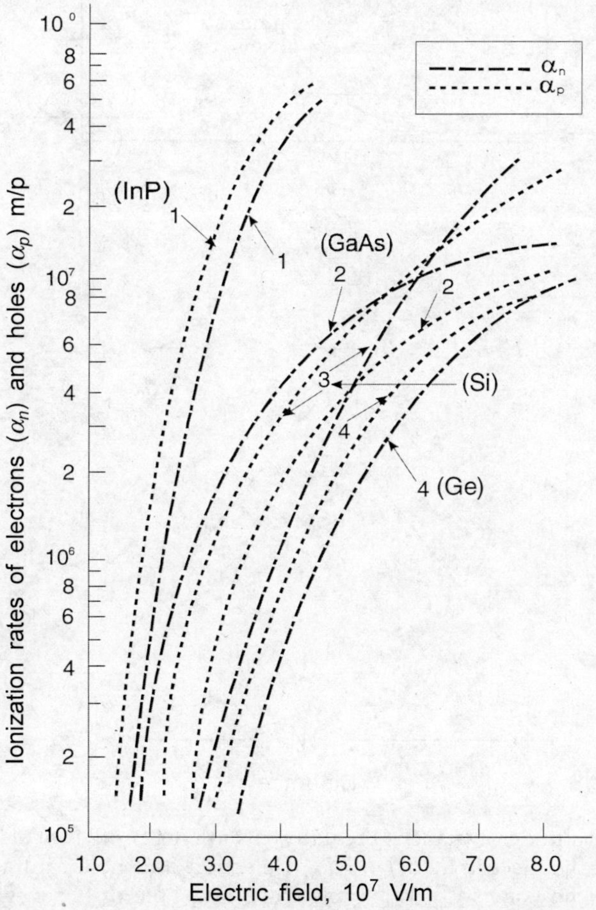

Figure 3.4 Ionization rate of electrons and holes in semiconductors versus electric field in: (1) InP, (2) GaAs, (3) Si, and (4) Ge.

Repeated multiplication of charge carriers takes place by impact ionization like an avalanche in the high field region around a junction biased to breakdown. When a small reverse saturation electron or hole current enters either edge of the depletion layer of a reverse biased junction, the current is infinitely multiplied by the high field around the junction and the junction undergoes avalanche breakdown as shown in Figure 3.5. The current multiplication factor is related to the ionization rates of charge carriers in the depletion layer [12].

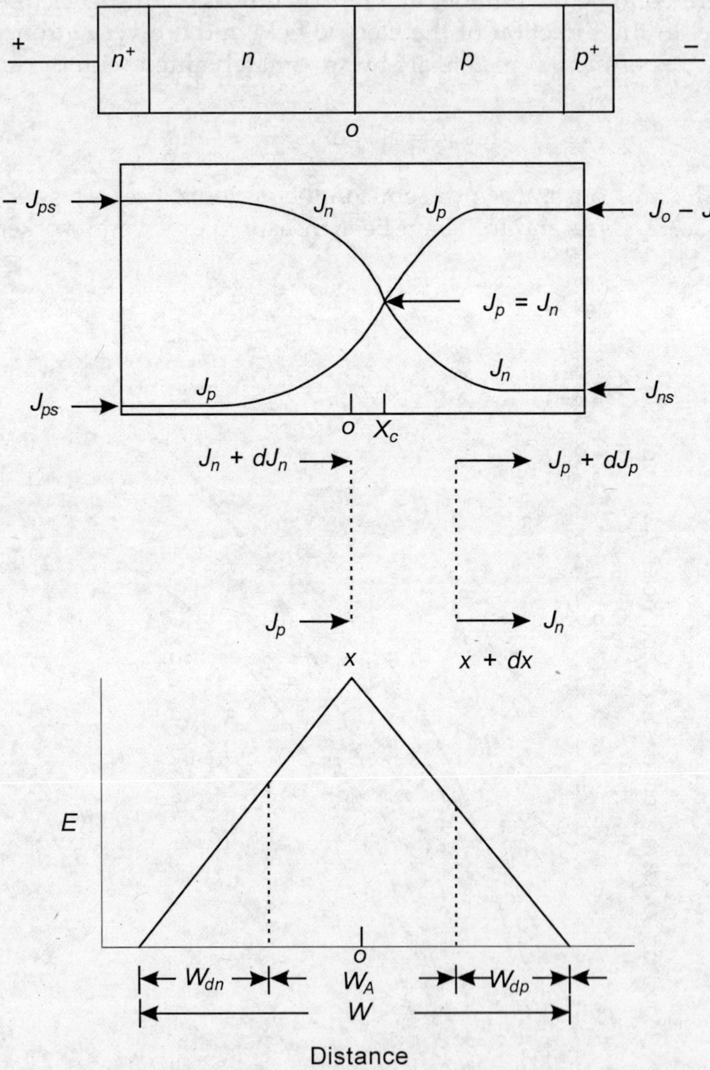

Figure 3.5 A p-n junction under avalanche breakdown showing profiles of electron and hole current density and the electric field $J_{ns} = J_{ps}$ = electron and hole saturation current densities; W_A = avalanche zone; W_{dn}, W_{dp} = electrons and hole drift zones, X_c = avalanche zone centre ($J_p = J_n$); W = depletion layer width.

$$\int_0^W \bar{\alpha}\, dx = 1 - (1/M) \tag{3.4}$$

$$\bar{\alpha} = \alpha_n \exp\left[-\int_0^x \left(\alpha_n - \alpha_p\right) dx\right] \tag{3.5}$$

where

W = Depletion layer width

M = Current multiplication factor

The current multiplication factor M is related with breakdown voltage by

$$M = 1/[1 - (V/V_b)^n] \tag{3.6}$$

Where V is the applied voltage, V_b is the breakdown voltage and n is a numerical factor depending on the doping of p^+n or n^+p region, which lies between 3 and 6.

The breakdown condition is reached when M tends to infinite, i.e., $V = V_b$ and is given by

$$\int_0^W \alpha\, dx = 1 \tag{3.7}$$

If $\alpha_n = \alpha_p = \alpha$, the breakdown condition is given by

$$\int_0^W \alpha\, dx = 1 \tag{3.8}$$

The variation of avalanche breakdown voltage V_b, depletion layer width W, maximum field at breakdown versus background doping concentration was studied and presented by Sze and G. Gibbons [13] for one sided abrupt junction in Ge, Si and GaAs. The results indicate sharp fall of V_b and W with increasing background doping.

The depletion layer of an IMPATT diode is commonly divided into a high field avalanche region of width W_α, where most of the multiplication takes place, and the lower field drift region, where the field is high enough for carriers to move with saturated velocity but not high enough for charge multiplication along their path. It is usual to define the avalanche zone width X_a by $\int \alpha\, dx = 0.95$. Thus 95% of the impact ionization takes place in avalanche zone. Charge carriers emerge in large numbers from the avalanche zone and interact with the field while crossing the drift zone. This interaction leads to the generation of high frequency oscillations. Thus the cavity should be tuned to get a resonant frequency as

$$f = 1/2\tau = v_s/2L \tag{3.9}$$

since the transit time τ should be equal to half the time period of oscillation.

Problem 3.1 An IMPATT diode has a pulsed operating voltage of 120 V with a pulsed operating current of 0.8 A. The efficiency is about 15%. Calculate (i) the output power and (ii) the duty cycle, if the pulsed width is 0.01 ns and the frequency is 16 GHz.

Solution: (i) The output power $P_{out} = \eta P_{in} = 0.15 \times (120 \times 0.8) = 14.4$ W.

(ii) The duty cycle = Pulse duration/Pulse repetition = $\tau/T = f = 16 \times 10^9 \times 0.01 \times 10^{-9}$ = 16%

Problem 3.2 An IMPATT diode has the following parameters:

Carrier drift velocity = 10^5 m/s
Length of the drift space = 5 μm

Calculate the frequency of oscillation produced.

Solution: From Equation (3.9) the frequency of oscillation

$$f = v_s/2L = 10^5/(2 \times 5 \times 10^{-6}) = 10^{10} = 10 \text{ GHz}$$

3.3 IMPATT STRUCTURES AND DOPING PROFILES

The doping profile of IMPATT diodes can be varied to produce different field distribution and corresponding avalanche and drift zones. The three basic types of IMPATT diodes are (i) single drift region (SDR), (ii) double drift region (DDR), and (iii) Double avalanche region (DAR) structure with flat doping profile in active regions. The SDR diode consists of a single avalanche zone and a single drift zone and was first proposed by Read with p^+nn^+ structure. A simple SDR is usually realized as p^+nn^+. The doping density for p^+ and n^+ is around 10^{20} atoms/cm^3 and the same for p and n regions is between 10^{16} and 10^{17}atoms/cm^3, and for i region it is around 10^{13} atoms/cm^3. A DDR diode has a p^+pnn^+ structure that consists of two drift layers, one for the electrons and other for holes on either side of the central avalanche zone. Electrons and holes emerge from the central avalanche zone and more respectively towards the positive and negative terminals of the device with saturated drift velocity. The DDR diode produces more high frequency power than the SDR, as each type of carriers in the corresponding drift zone makes the contribution to the power. A DAR diode has a p^+nipn^+ structure that consists of one drift zone sandwiched between two avalanche zones [14]. The electrons and holes from the two junctions travel across the central *i*-region in opposite directions and deliver power. Due to the cancellation of mobile space charge in the central drift region, the DAR has very little distortion of field profile in the presence of a large mobile space charge. Impurity charge bumps can be suitably introduced in the depletion layer by molecular beam epitaxy (MBE) or by ion implantation to produce high-efficiency low-high-low or high-low IMPATT diodes. The doping profiles are designed so that the resulting electric field profiles may constrict the avalanche zone. Thus one can have high-low or low-high-low IMPATT diodes with SDR or DDR structures where the avalanche zone can be suitably constricted by impurity bumps to realize high efficiency. The doping profiles of SDR and DDR structures with flat and low-high-low doping as well as for DAR and pin diodes are shown in Figures 3.6 and 3.7 respectively with the corresponding distribution of the electric field.

3.4 BASIC PRINCIPLE OF GENERATION OF MICROWAVES IN IMPATT DIODES

Microwave generation in an IMPATT diode can be explained using a simple SDR structure. In an SDR (Read or p^+nn^+ or n^+pp^+) the narrow avalanche zone situated at the junction end of the depletion layer is followed by a drift zone. If a sinusoidal electric field is applied to

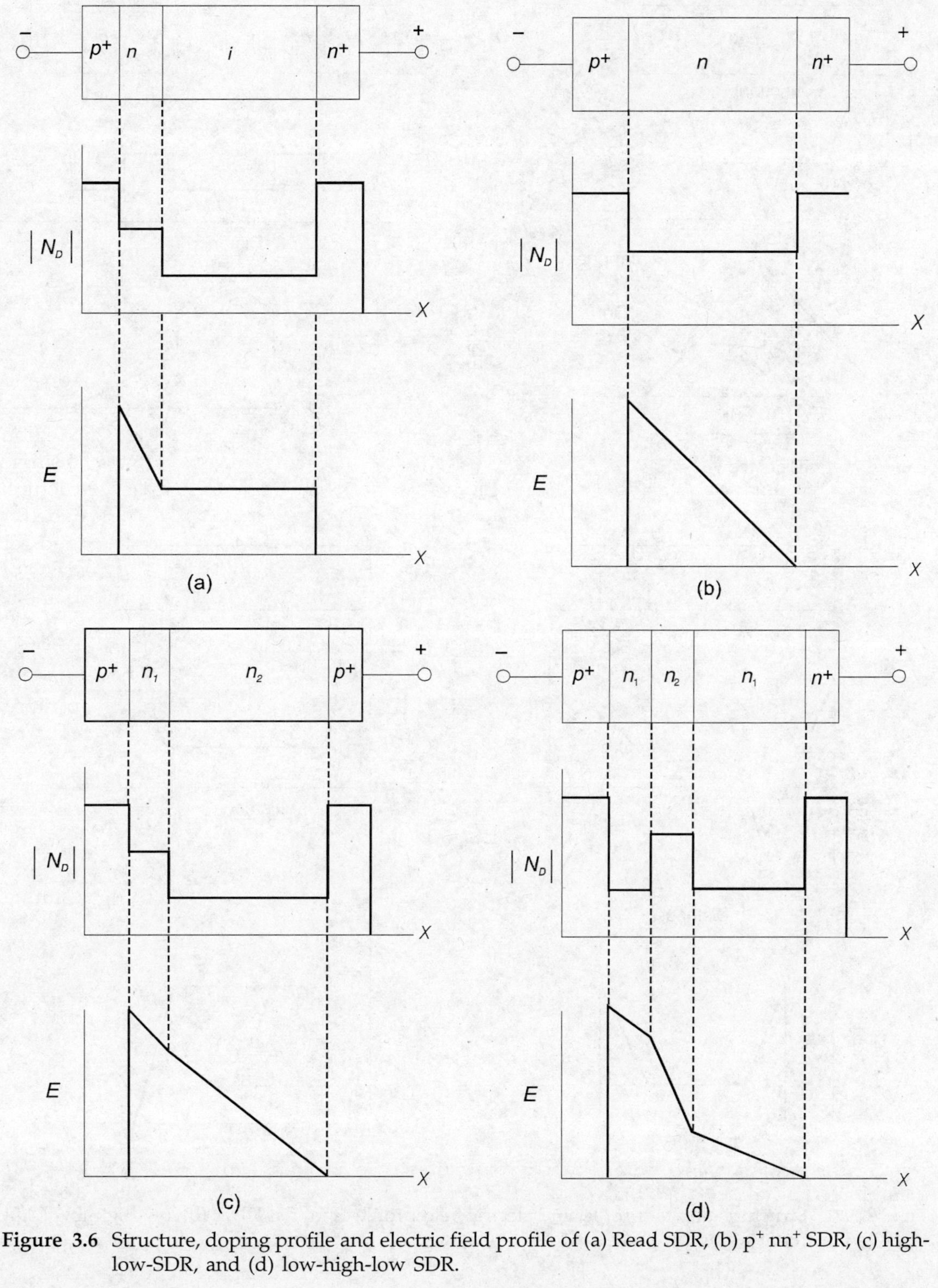

Figure 3.6 Structure, doping profile and electric field profile of (a) Read SDR, (b) $p^+ nn^+$ SDR, (c) high-low-SDR, and (d) low-high-low SDR.

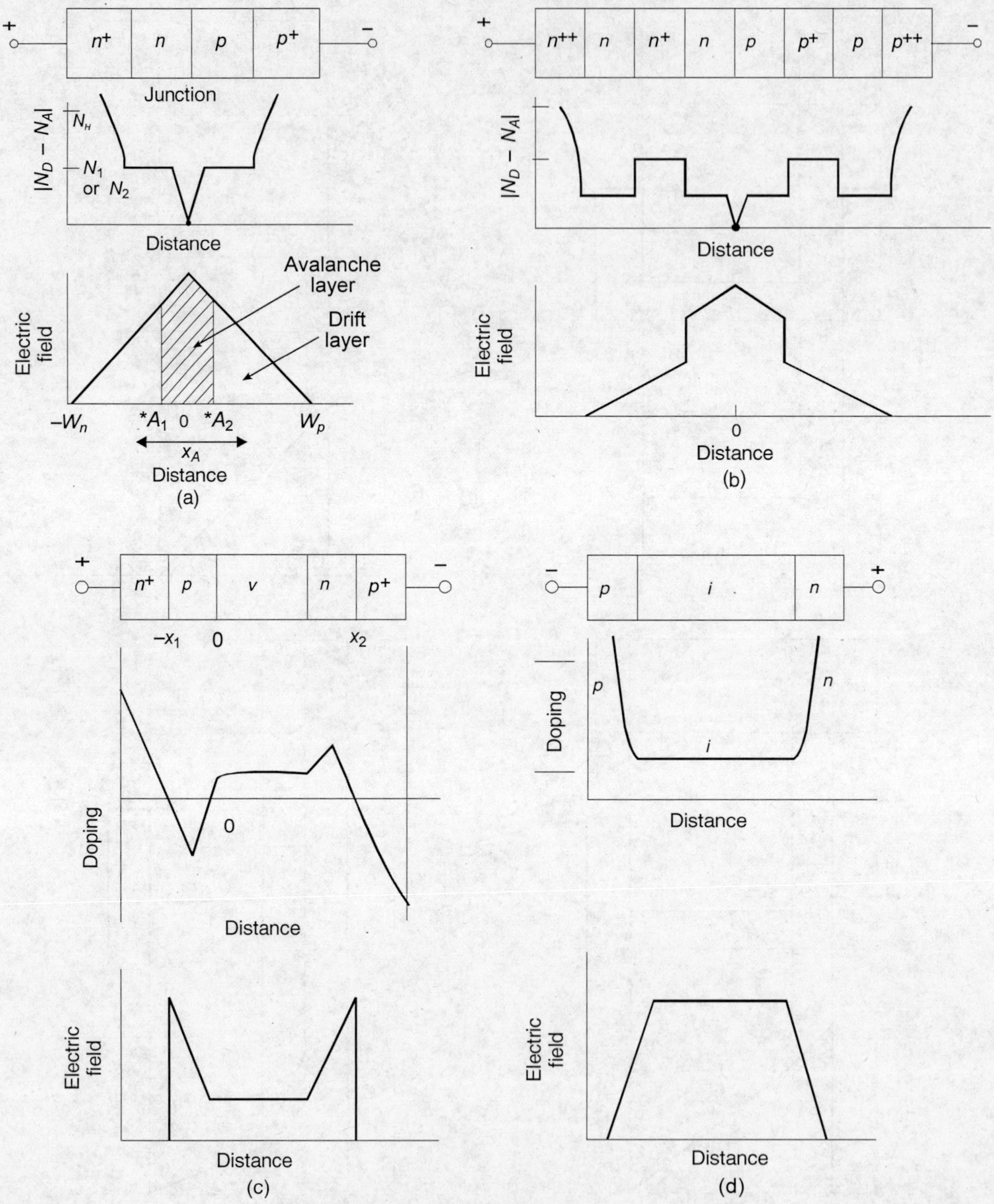

Figure 3.7 Structure, doping profile, and electric field profile of (a) flat DDR, (b) low-high-low DDR, (c) DAR, and (d) pin.

a device biased to the threshold of dc breakdown, an avalanche of e-h pairs created reaches its peak after the peak of the ac field has passed. This is because the number of e-h pairs created is proportional to the product of ionization rate of an individual carrier, which is highest at the instant of the peak field, and the number density of charge carriers present at that time. Since the number density goes on increasing as long as the applied field is added to the dc field, the peak of e-h pairs created is delayed with respect to the peak of ac field by an angle approximately equal to $\pi/2$. This delay is called the avalanche buildup time. The charge pulse thus formed is delayed by $\pi/2$ with respect to the peak of the ac field, which then crosses the drift zone with saturated velocity and produces a constant induced current in the external circuit during the time of transit w/v_s.

Thus one gets a positive rectangular current pulse when the ac field superimposed on the dc field is negative as shown in Figure 3.8. The fundamental component of this current is therefore in antiphase with the ac field, which causes a high frequency negative resistance.

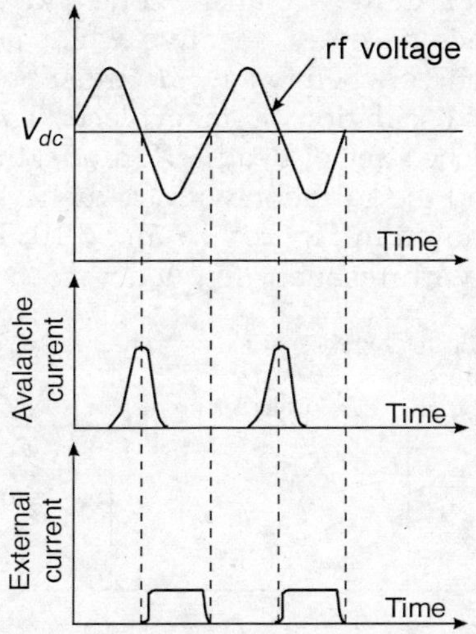

Figure 3.8 Waveforms of rf voltage, avalanche current and induced external current in a Read diode.

The avalanche build up time delay together with transit time delay leads to approximately 180° phase delay between the ac current and the applied ac voltage, leading to a high frequency negative resistance. The oscillation generated by IMPATT diodes in the microwave frequency range which can be understood by taking the example of a common silicon (p^+ nn^+ SDR) device with n-region of width w. The time required to cross width w with drift velocity v_s is w/v_s.

Thus half the time period of oscillation is given by w/v_s and the frequency is given by

$$f = v_s/2w$$

For a typical n-layer thickness of 5 μm and drift velocity of 10^5 m/s, we have

$$f = 10^5/2 \times 5 \times 10^{-6} \text{ Hz} = 10^{10} \text{ Hz} = 10 \text{ GHz}$$

Thus, in principle 100 GHz can be generated by simply taking epitaxial layer of 0.5 μm thickness.

3.5 SMALL SIGNAL ANALYSIS BASED ON ANALYTICAL APPROXIMATION

In order to understand the high frequency properties of IMPATT, Read undertook a small signal analysis of the SDR, which was further improved by Gilden and Hines [15]. For the sake of analytical convenience, the ionization rates and drift velocities were considered identical ($\alpha_n = \alpha_p, v_{ns} = v_{ps}$); and small sinusoidal variations of electron density, hole density and electric field were imposed on the dc quantities. The diode impedance is approximately given by

$$Z = w_d^2/[2Av_s\varepsilon_s\ (1 - \omega^2/\omega_\alpha^2)] + R_s + j/[\omega C\ (1/\omega_\alpha^2)/(\omega^2 - 1)] \qquad (3.10)$$

Where ω is the angular frequency, w_d is the width of the drift zone, A is the diode area, R_s is the parasitic positive series resistance due to substrate and undepleted epitaxial region, ε_s is the dielectric constant, $C = (\varepsilon_s A)/w$, the total depletion region capacitance and w = depletion region width, ω_α is the avalanche resonance frequency given by $\omega_\alpha = (2\alpha' v_s J_0/\varepsilon_s)^{1/2}$, where $\alpha' = \delta\alpha/\delta E$ in the avalanche zone and J_0 is the dc current density. This shows that the diode resistance becomes negative when $\omega > \omega_\alpha$. The diode reactance is inductive for $\omega < \omega_\alpha$ and capacitive when $\omega > \omega_\alpha$, R_s must be less than the magnitude of the active negative resistance for oscillation to take place. The schematic diagram and the equivalent circuit of the avalanche zone of Read SDR diode, the equivalent circuit of the Read diode for small transit angle and the variation with frequency of the real and imaginary parts of the active impedance are shown in Figures 3.9 and 3.10. The avalanche zone behaves as a parallel LC resonant circuit with resonant frequency ω_α, where

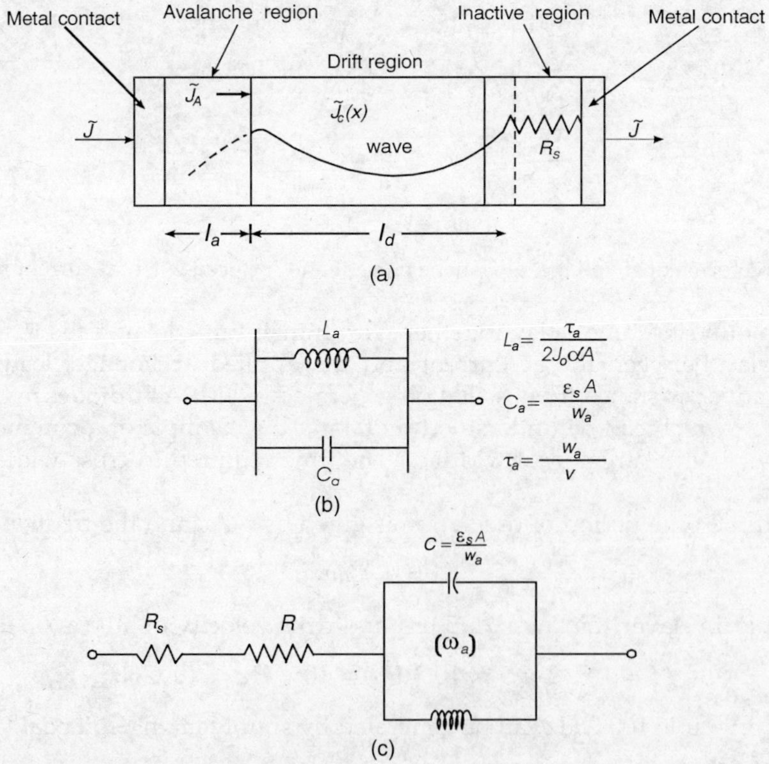

Figure 3.9 (a) Schematic diagram of avalanche, drift and inactive regions of a Read diode, (b) equivalent circuit of the avalanche zone of a Read diode, and (c) equivalent circuit of a Read diode for small transit angle.

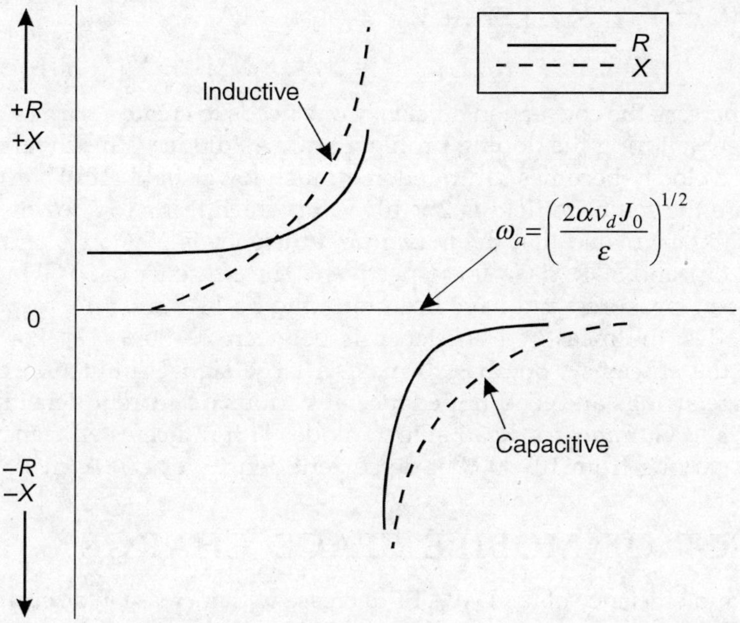

Figure 3.10 Resistance-reactance plot of a Read diode against frequency.

$$L_\alpha = \omega_\alpha / 2J_0 \alpha' A v_d$$

and

$$C_a = \varepsilon_s A / \omega_\alpha$$

3.6 DC TO MICROWAVE CONVERSION EFFICIENCY

The transit of carrier pulses in the drift region after avalanche build up leads to a rectangular waveform of the current whose fundamental component is in antiphase with the sinusoidal voltage across the device. For full modulation the fundamental component of the carrier current is given by $(4/\pi)I_0 \sin \omega t$. The dc to rf conversion efficiency of the device is thus given by the ratio of the output ac power to the input dc power

$$\eta = P_{ac}/P_{dc} = \int_0^{2\pi} [4I_0/\pi \sin \omega t \cdot V_m \sin \omega t \, dt]/I_0 V_0$$

$$= (2/\pi) \times V_m/V_0 \tag{3.11}$$

Where I_0 is the dc current, V_0 is the dc voltage and V_m is the ac voltage magnitude. The maximum field swing E_m allowed in drift region of silicon diode is assumed to be one half the dc drift field E_d, so that the electric field in the drift region does not fall below that required for velocity saturation, i.e.,

$$E_m = (1/2)E_d$$

The magnitude of the ac voltage is approximately given by

$$V_{ac} = (1/2)[(W_a + W_d)E_d]$$

The efficiency of silicon IMPATT diode for W_a small compared to W_d is given by Scharfetter and Gummel as

$$\eta = [(2/\pi) \times (1/2)V_D]/(V_A + V_D) = (1/\pi) \times V_D/(V_A + V_D)$$
$$= (1/\pi) \times [1/\{1 + (V_A/V_D)\}] \quad (3.12)$$

Thus to increase the conversion efficiency one needs to create a narrow avalanche region by appropriately tailoring the doping profile. For GaAs diodes, a much larger field swing is possible as the velocity becomes saturated at a much lower field (10^5 V/m) than in the case of silicon where the requisite field is 2×10^6 V/m which leads to a lower efficiency. It has been theoretically estimated that the maximum efficiency is 15%, 21% and 38% for X-band SDR (Si), DDR (Si), and SDR (n-GaAs) respectively. The efficiency of 50 GHz Si DDR and SDR diodes have been measured and have been found to be 14% and 10% respectively [16]. For GaAs SDR diodes, the measured efficiency is between 20–30% [17]. For a more realistic calculation of the efficiency, one has to make a large signal simulation of the device for different voltage swings and load impedance at various dc current densities. A calculation by Scharfetter and Gummel for Read silicon diodes [18] indicates efficiency variation from 5–15% for V_m varying from 10–30 V at dc current density of 200 A/m^2.

3.7 EFFECT OF MOBILE SPACE CHARGE

Power output and efficiency of the IMPATT decrease with increase in dc current density when the flow of mobile charge distorts the electric field profile and expands the avalanche zone. The effect of mobile space charge and sharp expansion of avalanche zone for silicon DDR has been discussed [19] and shown in Figure 3.11. The dc results indicate that field profile for flat profile diodes get distorted and remain high over a wider portion of the depletion layer as shown in the Figure 3.11 for X-band flat profile Si DDR.

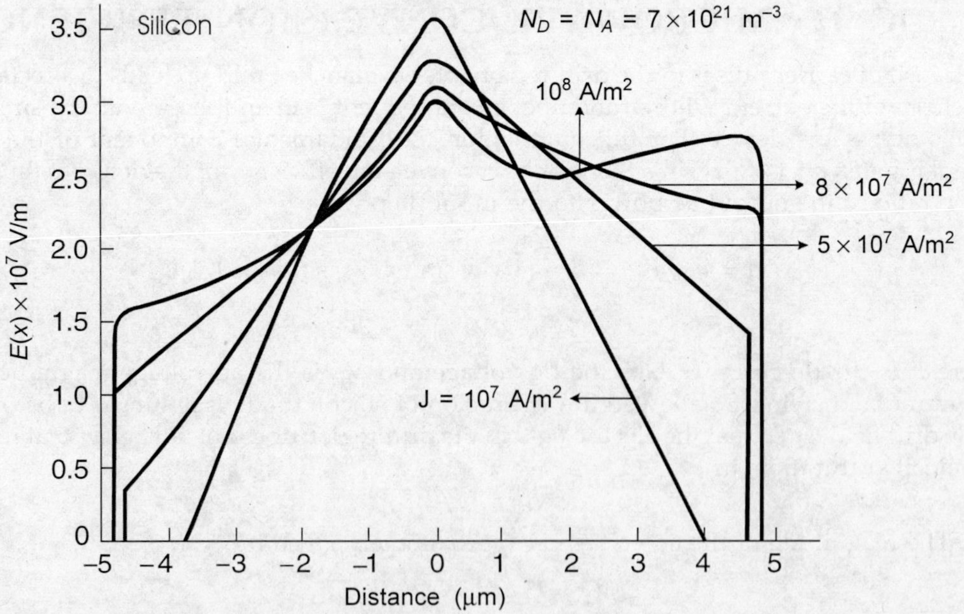

Figure 3.11 Electric field profiles of symmetrical double-drift region (n^+npp^+) silicon IMPATT for different values of dc current density.

This leads to an expansion of central avalanche zone from 26% of depletion layer at $J_0 = 10^7$ A/m^2 to 70% at the threshold of 10^8 A/m^2 when the avalanche zone sharply expands as indicated by the current profile in Figure 3.12. This leads to sharp deterioration of dc to microwave conversion efficiency because of the decrease in drift zone widths on the either side as indicated in Figure 3.13. The small signal G-B plot of X-band DDR has been drawn for various values of dc current density [20] and is shown in Figure 3.14. The small signal analysis indicates that the magnitude of the maximum negative conductance initially increases with dc current but begins to decrease when the dc current exceeds the threshold value for sharp expansion of the central avalanche zone.

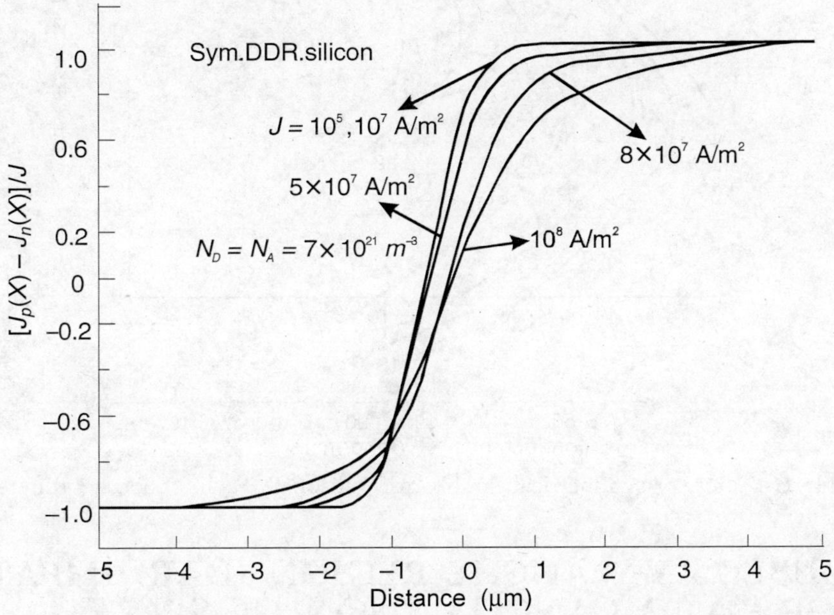

Figure 3.12 $[J_p(X) - J_n(X)]/J$ profiles of a symmetrical DDR silicon IMPATT for different values of dc current density.

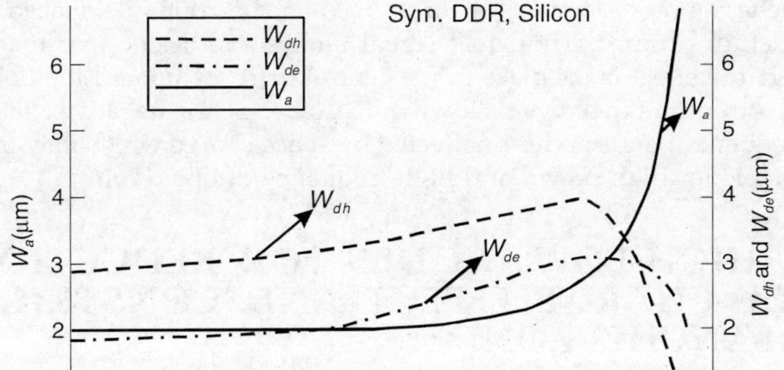

Figure 3.13 Plots of avalanche layer width W_a electron drift region width W_{de} and hole drift region width W_{dh} against dc current density.

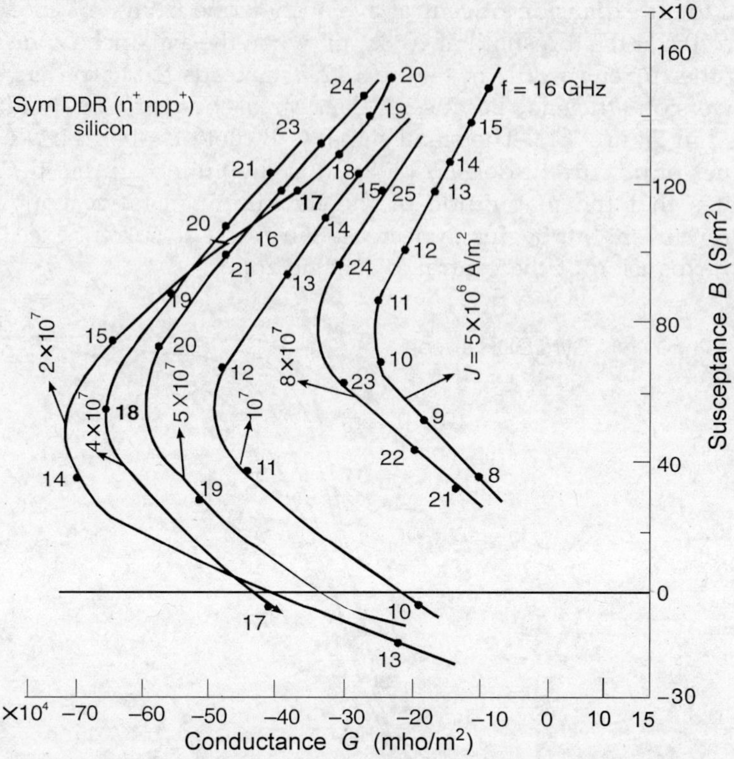

Figure 3.14 *G-B* plots of symmetrical double-drift silicon IMPATT diode for different values.

3.8 DOUBLE AVALANCHE REGION (DAR) IMPATT DIODE AND CANCELLATION OF SPACE CHARGE EFFECT

Double avalanche region diode consists of two-avalanche zone interspaced by a drift zone and has a structure p^+nipn^+. The small signal analysis indicates that the cancellation of mobile space charge effect takes place in the central drift region and the field profile is not affected at high dc current density as shown in Figure 3.15. The negative conductance occurs over several frequency bands and is unaffected by space charge which may make this device potentially suitable for high power and high frequency of operation.

3.9 LOW-HIGH-LOW DIODES FOR REDUCTION OF SPACE CHARGE EFFECT AND CONSTRICTION OF AVALANCHE ZONE

Low-high-low IMPATT diodes are emerging as more efficient devices than normal SDR and DDRs. In such diodes impurity bumps on high doping density are incorporated by MBE technique in appropriate positions of the flat profile DDR or SDR, which decreases with effect of mobile space charge on this field profile on the drift zone and constricts the width of the

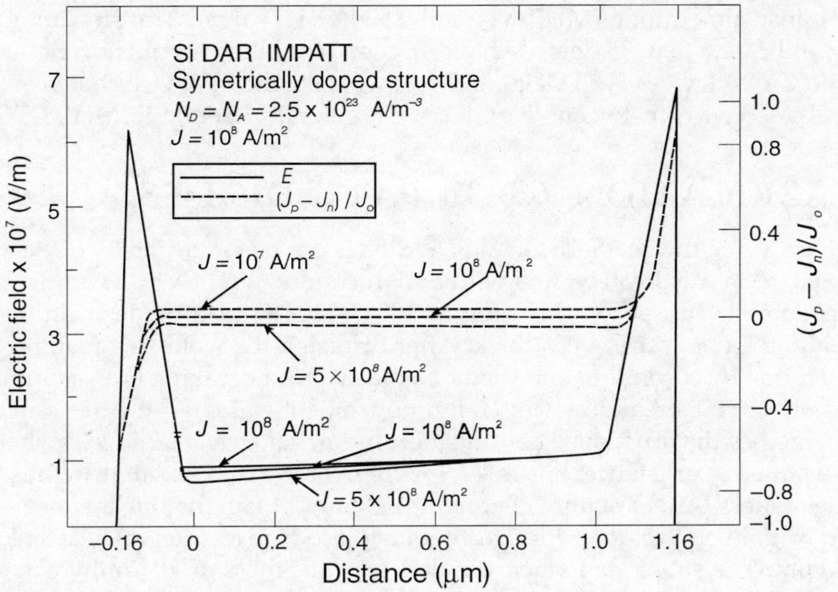

Figure 3.15 Field and carrier current profiles of symmetrically doped DAR IMPATT at various values of dc current density.

avalanche zone thereby increasing the efficiency of the device. Pulsed DLHL silicon devices have been shown to be capable of delivering large millimeter wave power at 94 GHz at dc current density of 10^9 A/m^2 [21].

3.10 HIGH FREQUENCY CONSIDERATION

As the frequency of IMPATT operation increases the width of the depletion layer decreases accompanied by a fall in breakdown voltage and power output. The maximum power output of an IMPATT diode falls sharply at high frequencies due to limitations regarding attainable impedance level in microwave circuit and the inherent limitations regarding the maximum electric field E_m. At breakdown and the saturation drift velocity v_s, the maximum input power P_m is related to frequency by

$$P_m f^2 = E_m \tag{3.13}$$

where $X_C = 1/2\pi fC$ and $C = \varepsilon_s A/w$ (the capacitance of the depletion region). Thus the maximum input power decreases as $1/f^2$, which indicates the electronic limitation of power output at millimeter wave frequencies. For the same circuit impedance level, the output power from an oscillator decreases with increases in frequency as $1/f^2$. Higher breakdown voltage at same frequency of operation provides higher power output, and thus DDR has advantage over SDR. InP with lower value of ionization rate than Si or GaAs for some electric field has higher breakdown voltage and would therefore be the suitable semiconductor for powerful millimeter wave source. But silicon remains the most important material for millimeter wave IMPATTs because of its advanced technology and stability. At high frequencies above 100 GHz, the maximum value sharply increases which in turn increases the possibility of tunnelling across the junction. Elta and Hadded [22] suggested mixed tunnelling avalanche transit time mode (MITTAT) and TUNNET (Tunnel Transit Time) mode. Their calculation indicated that Si devices have higher tunnelling frequency limit (~400 GHz) compared to GaAs devices (~75 GHz) and further that silicon devices between 400 and 500 GHz, and GaAs devices between 75 and 150 GHz would operate in MITTAT mode.

3.11 FABRICATION OF IMPATT DIODES

Lower frequency microwave (8–26 GHz) SDR diodes can be fabricated by diffusion technique [23]. Epitaxial n on n^+ silicon wafers with a high conductivity (0.003 Ω cm), arsenic doped substrate is subjected to boron diffusion at 1000°–1200°C to provide p^+nn^+ structure. For X-band diodes the epilayer is 4 μm thick. After diffusion the wafer is vacuum coated on the junction side, first by chromium and then by gold without breaking the vacuum. Chromium prevents short circuit breakdown by migration of gold atoms into the device at higher temperature across the junction. The transition in nn^+ layer has to be very sharp to reduce the positive parasitic resistance. For this a low temperature epitaxial growth is required for nn^+ epitaxial wafers. After vacuum coating at 10^{-8} mm of Hg, the thickness of metallic layer is increased by gold electro deposition technique to produce an integral heat sink of thickness around 200 μm. The substrate is then thinned to a thickness of 20 μm by chemical etching and then by mechanical polishing. The substrate side is then vacuum coated with chromium and gold. Photolithographic technique is applied to define diodes with 100 μm diameter on an integral heat sink of 500 μ square metre and the structure is called Mesa structure. Each device is then separated and encapsulated in a standard microwave package by using die bonding and wire bonding techniques. Fabrication process steps for X-band IMPATT diode have been described in Figure 3.16. IMPATT chips in the microwave frequency range (30 GHz) are packaged in standard microwave packages (S_4). The junction side of the chip is die bonded to a gold coated post of oxygen free high conductivity copper in a thermasonic bonding machine using heat, pressure and ultrasonic vibrations. The metal cap on the other side of the package is isolated from the post by a ceramic cylinder. The other side of the chip is wire bonded to the rim of the cap by means of gold wires. Diode packages are shown in Figure 3.17.

3.12 GaAs AND InP IMPATT DIODES

GaAs has emerged as a highly suitable semiconductor for IMPATTs in the lower microwave frequency range, where the efficiency of GaAs device has exceeded that of silicon devices. R. Goldwasser and R. Rosztcozy [24] reported high efficiency GaAs low-high-low SDR diodes delivering 2.9 W with 35.6% efficiency at X-band. Iglesius et al. [1] reported 12.5 W output at 6.1 GHz with 26.6% efficiency for SDR Schottky barrier high-low GaAs diodes. Alderstein et al. [25] reported 710 mW output for a GaAs Read diode with 9% efficiency in the 36–38 GHz range. Vasudev [26] reported 40 W of pulsed power from GaAs double drift IMPATTs at 9.4 GHz with 10% efficiency. At higher frequencies the power output of GaAs diodes is very low. Chang et al. [27] reported fabrication of a Read type GaAs IMPATT diode by VPE and obtained CW power of 5 mW with 0.5% conversion efficiency at 130 GHz. It can be stated that GaAs high-low and low-high-low diodes exhibit higher efficiency than any other IMPATT below 30 GHz.

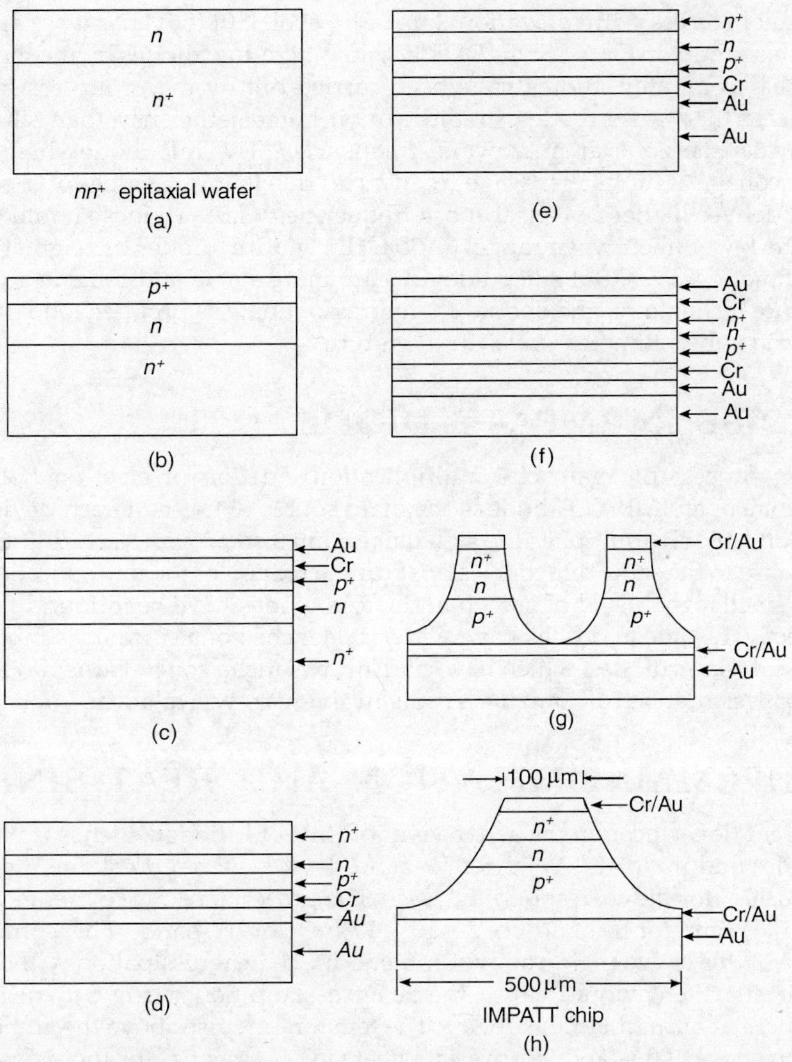

Figure 3.16 Fabricational process steps of X-band Si IMPATT diode: (a) Epitaxial wafer; (b) after boron diffusion; (c) after vacuum coating by Cr and Au; (d) after electroplating of Au; (e) after thinning of substrate; (f) after vacuum coating of Cr and Au on n^+ side; (g) after photolithography and mesa etching; (h) after chip separation.

The fabrication of ion-implanted single drift flat profile of p^+nn^+ InP IMPATT diodes operating at X-band was reported by Berenz et al. [16] in 1978. Singly ionized Be ions with 50 KeV energy were implanted into n-type InP epitaxial layers with sulphur as the chemical impurity on a tin-doped (100) oriented substrate grown by vapour phase epitaxy. The diodes exhibited CW output power of 1.6 W with 11% efficiency at 9.78 GHz. In low-duty-cycle pulse operation 6.1 W peak output powers was achieved at 10.8 GHz with 13.7% efficiency.

Computer studies with ionization data of Ito et al. [10] for GaAs was carried out by the author and his group on GaAs, SDR, DDR and low-high-low diodes by the methods outlined earlier. Similar computer studies have been carried out by the group for InP diodes. The results indicate that GaAs diodes should exhibit higher efficiency than silicon due to the narrower avalanche zone at microwave frequencies. The InP diode should have higher breakdown voltage at millimeter wave frequencies due to lower values of the ionization rate and should deliver higher power at those frequencies. GaAs diodes should surpass silicon devices in the lower microwave range (6–30 GHz) and InP diodes have great promise in the field of millimeter waves. But silicon devices are more stable and reliable and are based on a more mature technology, and should therefore dominate as premier solid-state sources for millimeter and sub-millimeter waves in the future.

3.13 NOISE IN IMPATT DIODE

The random process of avalanche multiplication creation of electron hole pairs in the avalanche zone of an IMPATT diode is the main source of noise in such devices. Avalanche multiplication not only multiplies the signal noise but also provides an additional substantial amount of noise to the same due to the statistical fluctuation of the multiplication factor. Noise measure at small signal level of silicon and GaAs diodes have been found to be 40 dB and 25 dB respectively. Silicon diodes have wider avalanche zone since α_n is much greater than α_p compared to GaAs diodes, which have narrow avalanche zones. GaAs diodes are thus less noisy due to less number of ionizing events in the narrow avalanche zone.

3.14 THERMAL LIMITATION AND HEAT SINK

A large current flows through the active layer of IMPATT diode which is up to 10^7 A/m^2 for X-band diodes and up to 10^9 A/m^2 for W-band devices with a breakdown voltage ranging from 80 V to 20 V for the corresponding cases. The power density is extremely high, ranging from 8×10^8 W/m^3 for X-band to 2×10^{10} W/m^3 for W-band. The continuous rise in temperature of this active region above ambient due to heat dissipation without appropriate heat sink arrangement would cause the reverse saturation current to rise exponentially leading to thermal burn-out of the device. The temperature rise above the ambient should not exceed much above 100°C and appropriate heat sink is essential for these devices. The most commonly used technique is to have an electroplated gold layer in the junction side of the wafer of thickness between 20 to 100 μm. The IMPATT chip is bonded to a gold-coated copper post of the package by thermo compression bonding for microwave diodes. For millimeter wave diodes, where heat dissipation is much larger copper heat sink is replaced by Type II (a) diamond which has a much larger (5 times) thermal conductivity than copper at room temperature. The millimeter package is therefore provided with diamond heat sink to which the device chip is bonded by a bonding machine.

3.15 MILLIMETER WAVE IMPATT PACKAGING

Between 30 and 100 GHz the device chip is usually bonded to a selectively metallized diamond heat sink which has a high thermal conductivity for flow of heat from the junction to the ambient and the same is sealed in a quartz ring with a number of gold ribbons bonded to the device and the top metal cap as shown in Figure 3.17. At a frequency higher than 100 GHz, a single and double metallized quartz stand-off mounting arrangement is used as shown in Figure 3.17(a) and Figure 3.17(b) to reduce the shunt capacitance to 0.01 pF, which is sufficiently small for operation up to 200 GHz. A direct contact method in which a thin bias pin makes contact with diode chip has been adopted at 100 GHz. At higher frequencies a direct contact has been adopted in Figure 3.18, which makes use of a wire welded on the tip of bias pin to contact the IMPATT chip. Copper and Diamond are both used as heat sinks for millimeter wave chips.

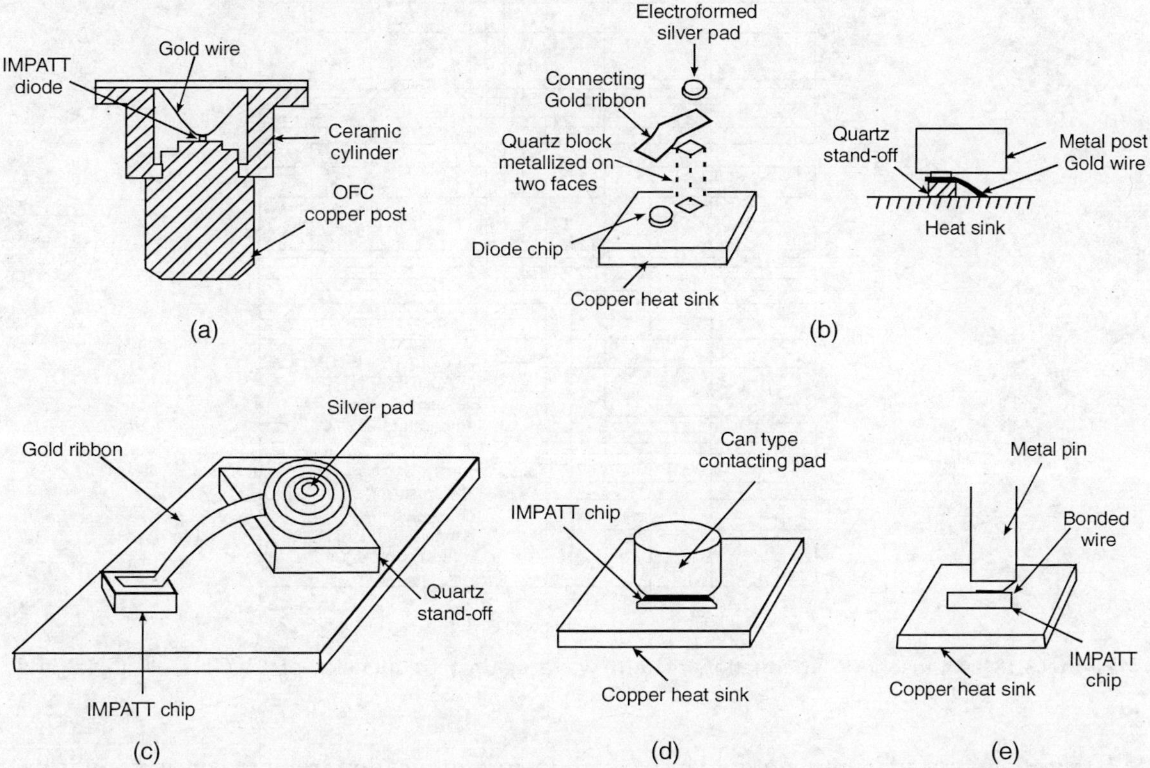

Figure 3.17 Packaging and bonding arrangements for IMPATT diodes: (a) microwave diode up to 30 GHz; (b)–(e) millimeter-wave diodes above 100 GHz.

3.16 MILLIMETER WAVE IMPATT DEVICES AND SOURCES

Millimeter waves cover the wavelength range from 10 mm to 1 mm corresponding to the frequency range of 30–300 GHz. In 1896 Sir J.C. Bose successfully generated, electromagnetic waves of wavelength 6 mm and frequency 50 GHz in Calcutta. He showed that these waves have the optical properties of the light waves. In recent years, due to the availability and emergence of semiconductor mm wave sources, mm waves are finding increasing

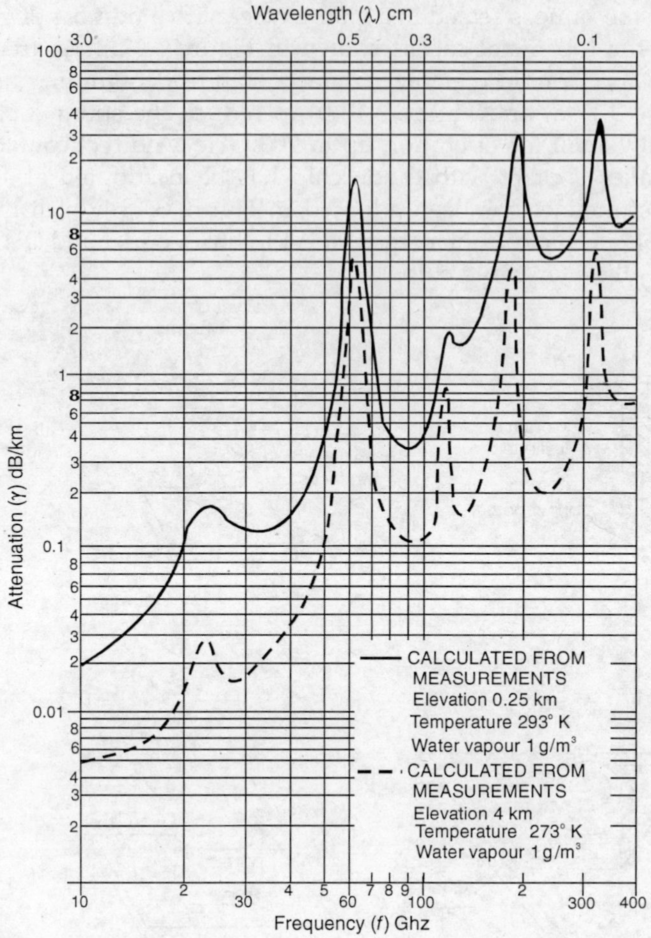

Figure 3.18 Atmospheric attenuation of mm wave against frequencies of 35 GHz, 94 GHz, and 140 GHz.

applications in radar, missile guidance, satellite communication, remote sensing, collision avoidance systems, radio astronomy and medical technology etc. The wide spectrum of applications is making the mm wave system development one of the most advanced technologies of Radio science.

Due to the ever increasing demand of communication, the microwave frequency ranges from 3–30 GHz (wave length range of 10–1 cm) are becoming saturated with land based and satellite based communication system. Millimeter waves undergo considerable attenuation while propagating in the atmosphere and communication is generally carried out around low atmospheric attenuation window frequencies. The atmospheric attenuation for mm waves against frequency is shown in Figure 3.18. IMPATT diodes have emerged as the most powerful solid state sources of mm wave electromagnetic power by the end of twentieth century and would be having increasing application in satellite based and land based communication systems. IMPATT diodes have been developed for low atmospheric window frequency bands around 35 GHz, 94 GHz, 140 GHz and 220 GHz. IMPATT diodes based on silicon are most suitable for generation of mm wave power. CW mm wave sources are mainly used for communication and the pulsed mm wave sources are mainly used for radar application and they have different design considerations. In this Chapter we will first discuss the CW IMPATT diodes and sources, which will be followed by pulsed mm wave IMPATT sources.

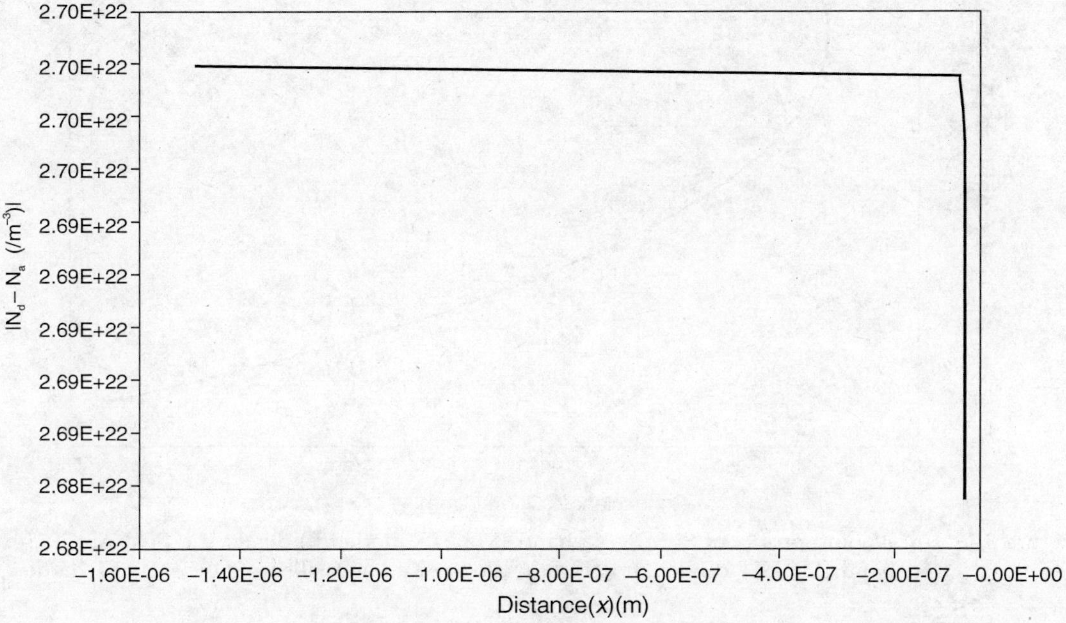

(a)

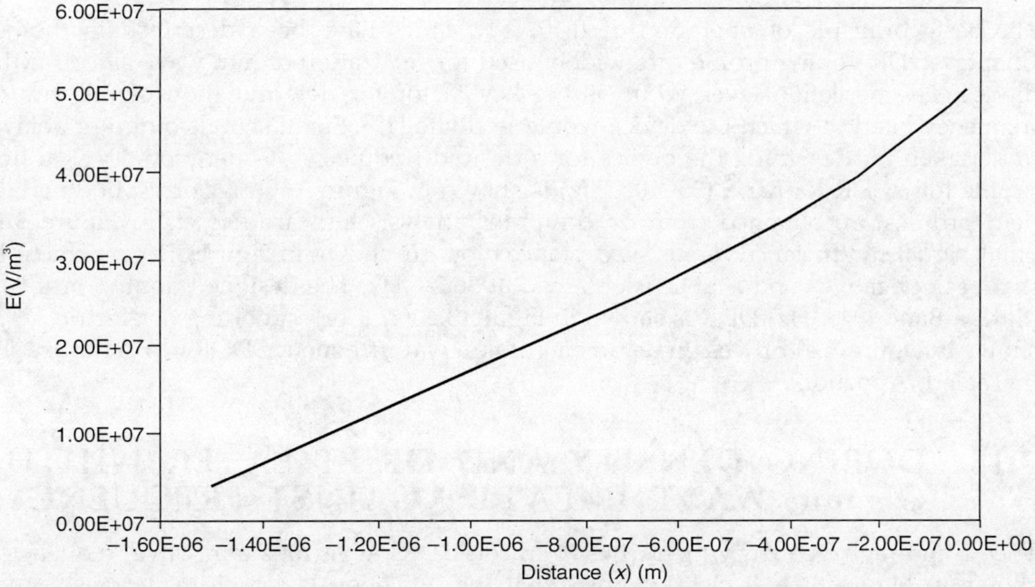

(b)

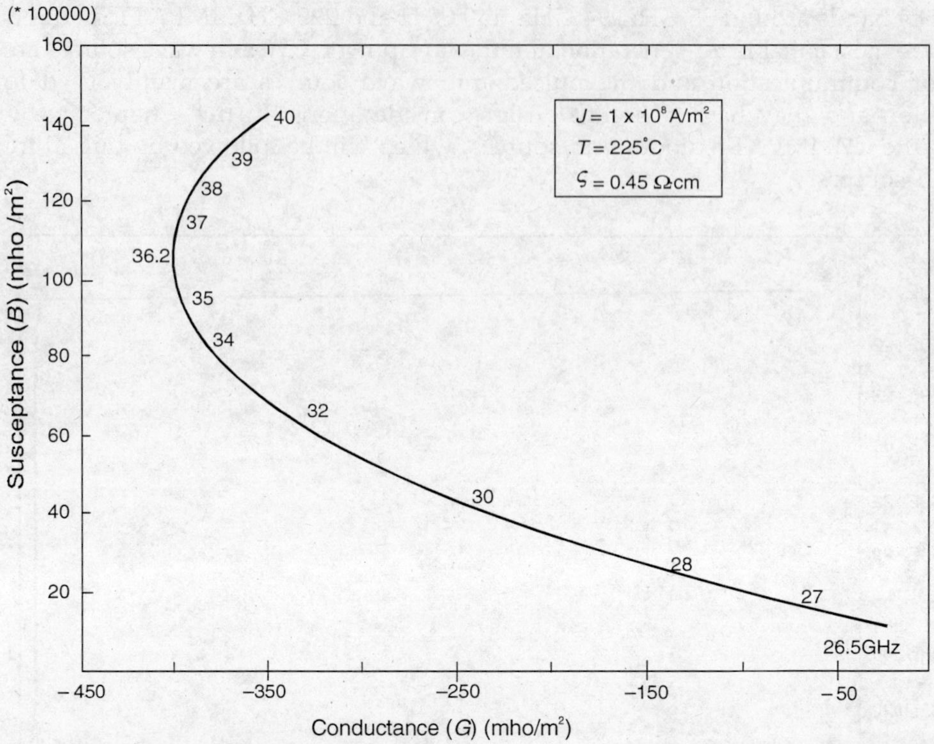

Figure 3.19 (a) Doping profile of Si SDR Ka-band IMPATT diode, (b) dc electric field profile for Si Ka-band IMPATT diode; and (c) *G–B* plot of Ka-band silicon SDR IMPATT diode.

3.17 CW mm WAVE IMPATT DIODES

The basic principle of operation of IMPATT diodes have been described in the earlier Chapter. DDR doping profiles are widely used for realization of mm wave silicon IMPATT diodes. The depletion layer width is the key factor for determination of the mm wave frequency band in which the device would oscillate [17]. Small signal computer analysis is undertaken for designing the device for a desired frequency of operation. Typical doping profile for a SDR Ka-Band (35–40 GHz) is shown in Figure 3.19(a). Corresponding electric field profiles are obtained from dc computer analysis and are shown in Figure 3.19(b). Small signal conductance versus susceptance plots are shown in Figure 3.19(c), which shows that peak negative conductance is obtained at 36.2 GHz. The designed doping profile for a flat Ka-Band (94 GHz) DDR is shown in Figure 3.20(a). Corresponding dc electric profile is shown in Figure 3.20(b). Design parameter of mm wave frequency DDR IMPATTs are shown in Table I, Appendix.

3.18 DOPING DENSITY AND DRIFT LAYER WIDTHS OF mm WAVE IMPATT AGAINST FREQUENCY

The dc and high frequency properties of silicon DDRs at various mm wave frequencies are shown in Figure 3.21. It clearly shows that the depletion layer width decreases and the

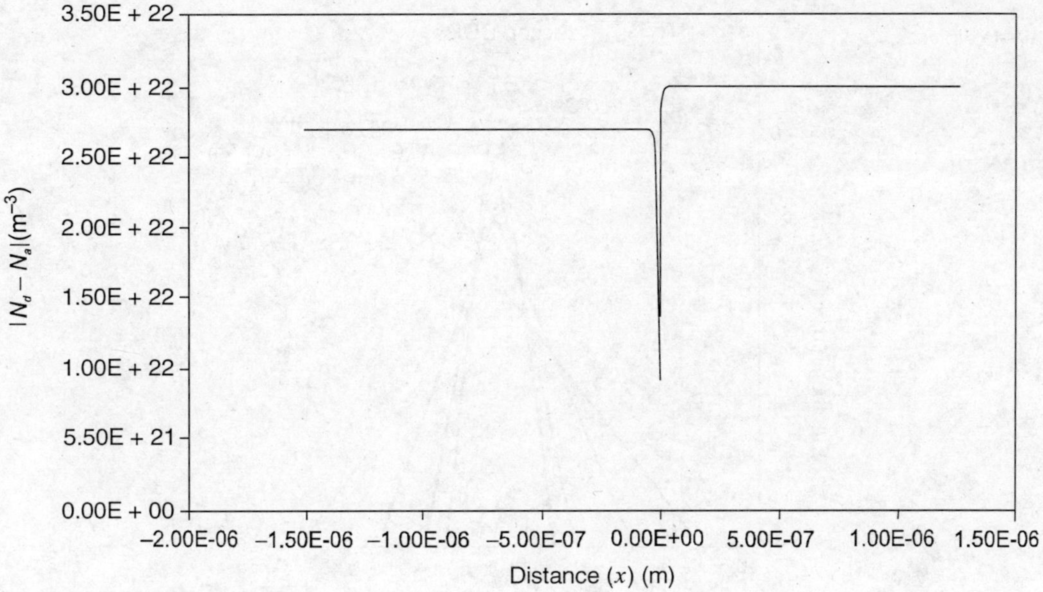

(a)

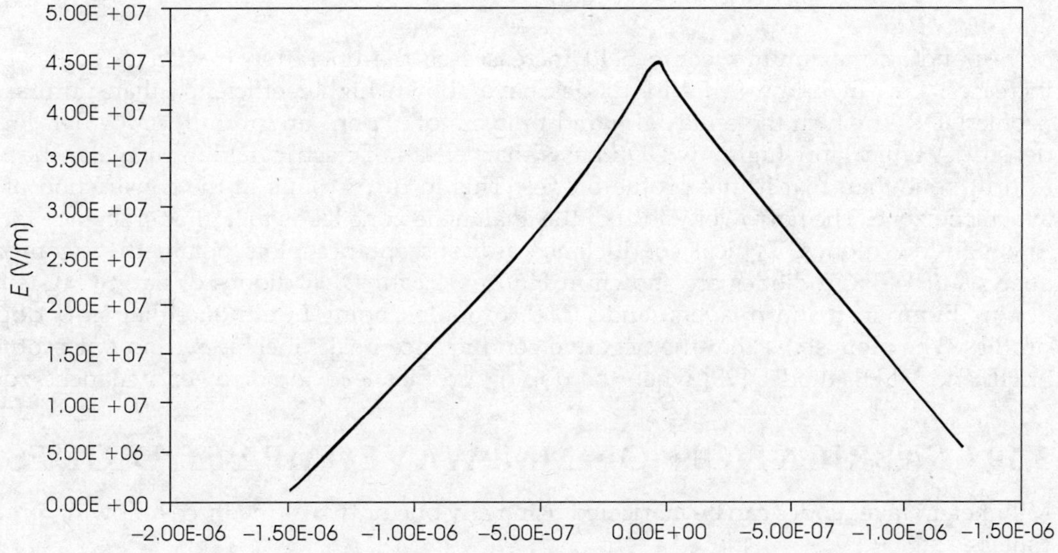

(b)

Figure 3.20 (a) Doping profile of Si Ka-band DDR-IMPATT diode, (b) Electric field profile for Si DDR Ka-band IMPATT diode.

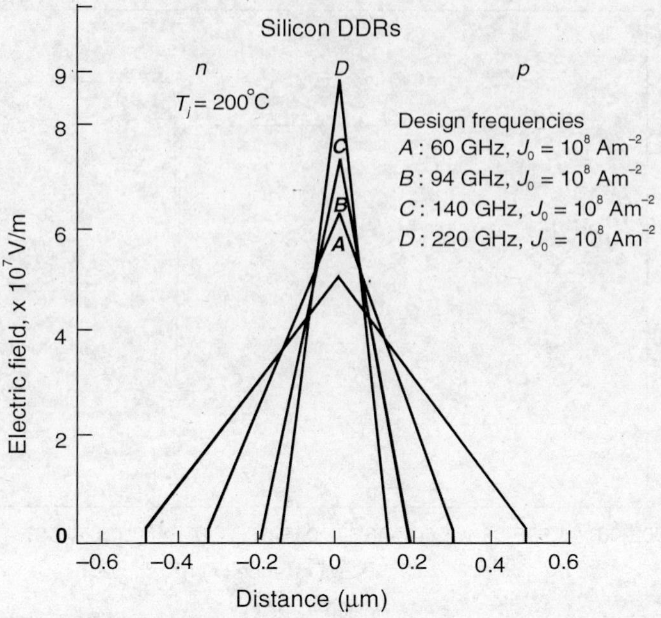

Figure 3.21 DC field profile of DDR-IMPATT diode at different walls and window frequencies.

corresponding maximum electric field increases as the operation frequency of the DDR increases. Low-high-low IMPATT diodes have shown higher efficiency than normal flat profile DDR in which the epitaxial n and p-layers of p^+pnn^+ are of uniform or flat doping density. A typical low-high-low DDR has a sharper fall of electric field from avalanche zone to drift zone than that in flat profile diodes. This in turn results in the constriction of the avalanche zone. The narrower width of the avalanche zone leads to higher efficiency for the low-high-low diodes. Typical conductance and susceptance plots obtained by computer analysis of W-Band diodes are shown in Figure 3.22 for DDR diodes of flat, SLHL (single doping bump in n and p-layers) and DLHL (double doping bump on either side) doping profiles. The plots show that the negative conductance peaks increases from flat profile to SLHL and DLHL diodes [28] when the doping profiles lead to narrower avalanche zones.

3.19 FABRICATION OF MM WAVE IMPATT DIODES

Millimeter wave diodes can be fabricated using any one or two of the three following process routes.

3.19.1 Diffusion

SDR mm wave diode junctions of either p^+nn^+ or n^+pp^+ can be formed using epitaxial nn^+ or pp^+ silicon wafers. After predeposition of boron on the epitaxial side of the wafer, the boron atoms are driven inside the epitaxial layer by raising the temperature of the furnace and increasing the time of drive in diffusion. The temperature and the time are appropriately adjusted according to the starting width of the epitaxial layer to realize the required junction

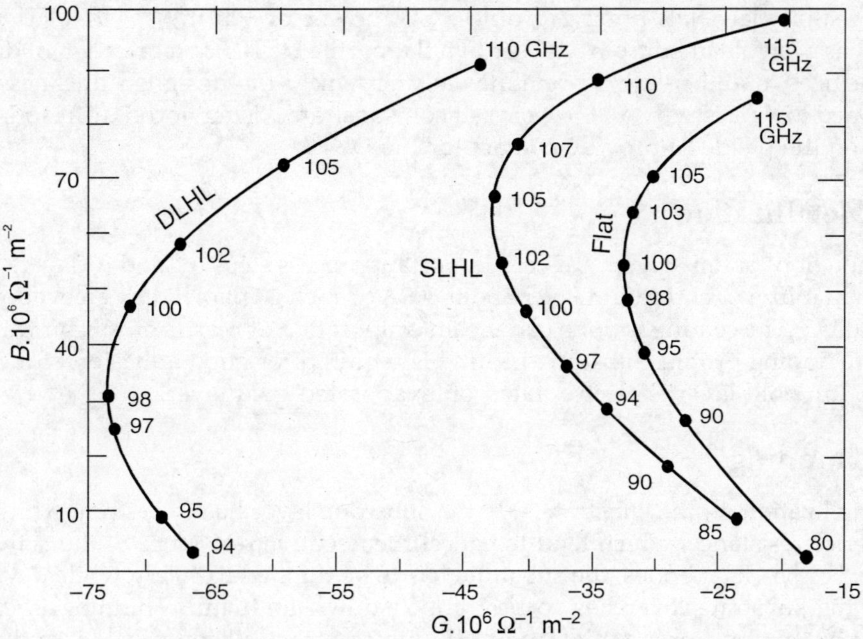

Figure 3.22 Conductance–Susceptance plot of W-band DDR diode for flat, SLHL, DLHL structures.

depths at which, boron concentration equals phosphorus and other Group V concentration in the epitaxial layer. This may be measured using a secondary ion mass spectrometer, which gives accurately the profiles of various impurities in the diffused p^+nn^+ silicon wafer. The technology for formation of SDR mm wave diodes by diffusion is simple and has been used to realize very high frequencies of operation, i.e., 35 GHz, 60 GHz, 94 GHz, 140 GHz, 220 GHz etc., although SDR gives lower power than DDR and low-high-low structures. An approximate formula for the frequency generation in a p^+nn^+ silicon diode of an n-layer width w is given by $f = 42/w$ GHz when w is in microns.

3.19.2 Ion implantation

Ion implantation provides a technique for obtaining accurately controlled impurity ion doping concentration with required penetration depths and slope of the doping profile. Double drift $p^+p\ nn^+$ device can be fabricated by ion implantation when the nn^+ wafer is bombarded with highly energetic trivalent impurity ions. DDR profile is obtained by implantation of boron into the n-type epitaxial surface layer of an nn^+ wafer. The boron flux is adjusted so that the boron concentration is twice that of the doner level i.e. phosphorus in the epitaxial layer. In this way the boron density in the p-region becomes equal to the phosphorus density in the n-region. The p^+ surface may be formed by shallow boron diffusion or by a high dose boron implantation at low energy.

3.19.3 Molecular beam epitaxy

In recent years the molecular beam epitaxy (MBE) technique is being increasingly used for IMPATT fabrication. By this technique very thin layers of appropriate doping density can be grown to realize flat DDR doping profiles. CW power of 600 mW at 94 GHz with 6.7% efficiency has been obtained from MBE grown flat profile DDR. For fabrication of double drift p^+p nn^+ devices, n and p-layers are epitaxially grown to the designed thickness on an n^+ arsenic doped low resistivity (0.03 Ω cm) silicon substrate. Using boron diffusion, a p^+ layer is formed on the p-side to provide contact for the DDR.

3.19.4 Metallization

Metallization of p^+ and n^+-layers are very important process steps. p^+ and n^+-layers are coated first with a titanium or chromium layer about 600A° which is followed by an evaporated gold layer of 3000A°. The coating is done by e-beam evaporation of metals on silicon wafer having appropriate doping profile. Substrate heating is required for good adhesion of metal layers. Finally a 5 μm gold layer is electroplated on evaporated gold layer.

3.19.5 Thinning

Before metallization on the substrate side the substrate layer has to be thinned to minimize positive series resistance, which tend to cancel the negative resistance of the active layer of the device. For 36 GHz diodes, the substrate has to be thinned to 40 μm while devices above 200 GHz, the substrate layer has to be thinned down to 10 μm. Thinning is carried out chemically by an appropriate mixture of HF (which etches silicon dioxide) and nitric acid (which oxidizes the Si) and acetic acid (which control the etch rate).

3.19.6 Photolithography and mesa etching

The diameter of the IMPATT chip would decrease with increasing operating frequencies and would vary from 100 μm at X-Band (10 GHz) to 50 μm at Ka-Band (35 GHz) and to 25 μm at W-Band (94 GHz) to maintain the optimum impedance level. After photolithography with appropriate masks and mesa etching the chips are separated by a diamond scriber for package and bonding.

3.19.7 Package and bonding of mm wave IMPATT chip

The packaging should provide a low thermal resistance between the diode chip and wave-guide mount and should be mechanically rugged and hermetically sealed. Thus the mm wave diode package differs from lower frequency microwave package in the sense that the heat generated due to electrical parasitic and rf looses becomes important at mm wave frequencies. A typical mm wave package for 30–100 GHz is shown in Figure 3.17. At very high frequencies (> 200 GHz) open packages comprising of Quartz stand-off are used to minimize parasitic capacitance (Figure 3.17). Figure of millimeter wave diode package of IMPATT diode is given in Figure 3.23.

3.20 mm WAVE IMPATT OSCILLATOR CIRCUITS AND PERFORMANCES

The packaged IMPATT diode is conveniently placed in a wave-guide for extracting oscillatory power. Resonant cap circuit with a back short tuner provides a simple arrangement for

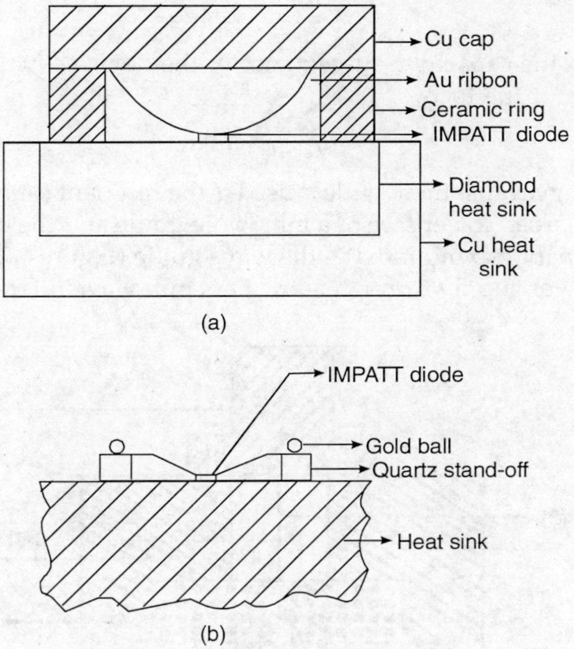

Figure 3.23 Millimeter wave diode package.

a mm wave oscillator and is widely used. An integrated heat sink cum wave-guide mount in which the diode is placed between a circular metal cap (resonant cap) of appropriate diameter and broad face of the wave-guide has been found to be suitable and will be described below.

3.21 CW MILLIMETER WAVE OSCILLATORS

We will first discuss how mm wave power can be obtained from a mm wave circuit using a mm wave IMPATT diode as the active device. The diode having device admittance $(G_D + jB_D)$ is mounted in a microwave/millimeter wave cavity and is biased to breakdown and is connected to load having admittance $G_L + jB_L$ (Figure 3.24). The condition for stable oscillation is fulfilled when the magnitude of the negative conductance of the device equals the magnitude of the positive conductance of the load,

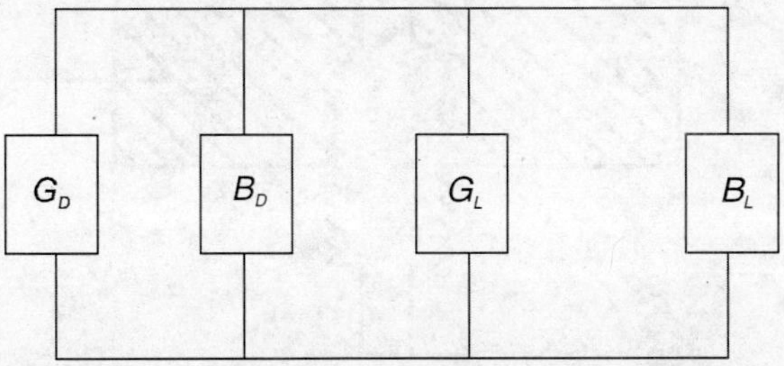

Figure 3.24 Equivalent circuit of IMPATT oscillator.

$$|G_D| = |G_L|$$

and the magnitude of the capacitive susceptance of the device equals the magnitude of the inductive susceptance of the load.

$$|B_D| = |B_L|$$

A convenient mounting arrangement widely used is the resonant cap mount when the device is placed between the broad lower face of a mm wave-guide and the bottom face of a circular metal disc called resonant cap mounted in the wave guide (Figure 3.25). The bias is given by a constant current power supply through a low pass mm wave filter connected with the top

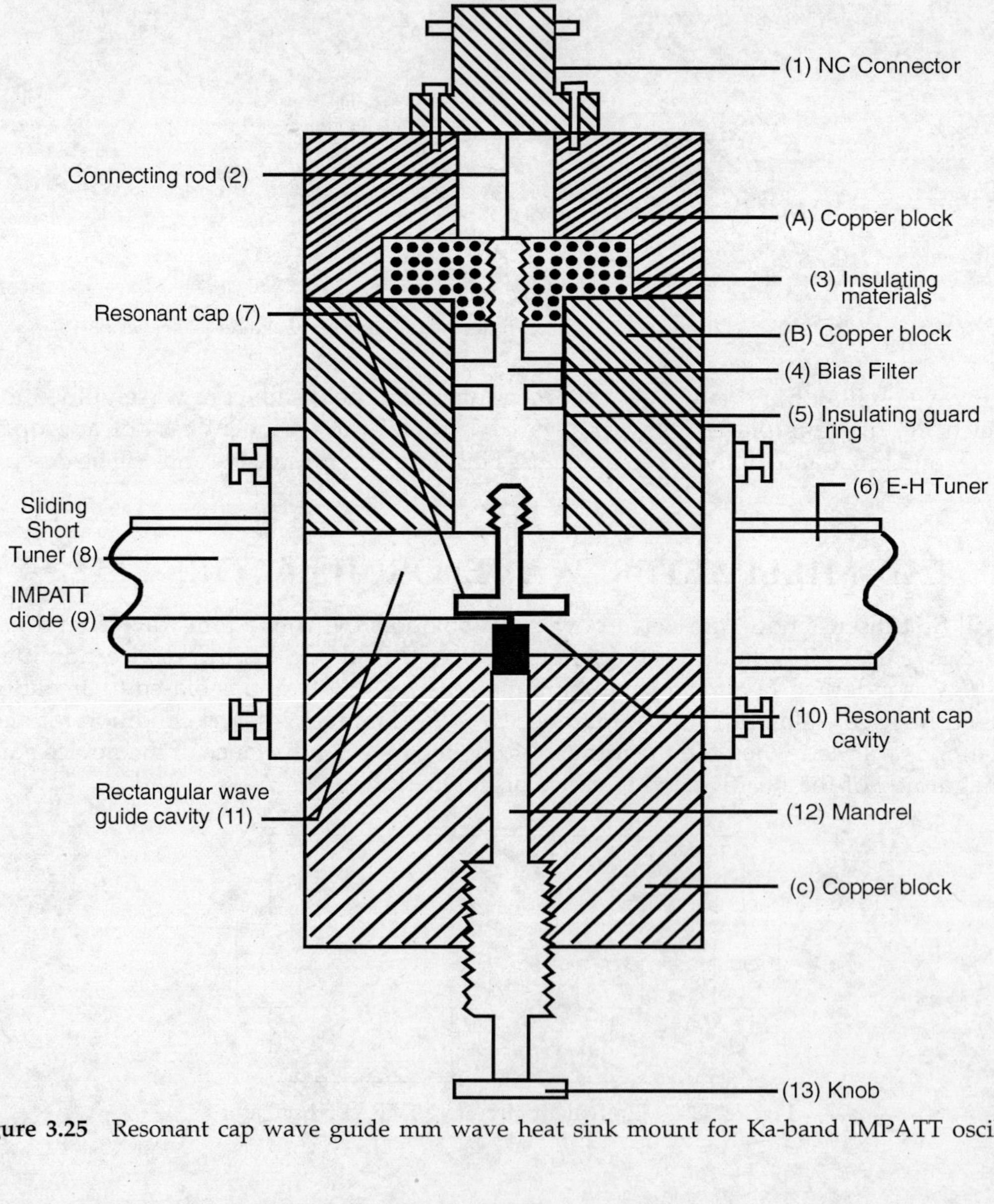

Figure 3.25 Resonant cap wave guide mm wave heat sink mount for Ka-band IMPATT oscillator.

surface of the resonant cap. The low pass π type filter prevents high frequency oscillations from coupling with the power supply when dc or low frequency bias current can pass through the filter and the diode. Considerable heat is generated in the diode junction where large current flows at breakdown. Therefore an integral heat sink cum wave-guide cavity is used which is shown in Figure 3.25. The resonant cap wave guide mount consists of a resonant cap structure under which the diode is embedded and the two together are mounted in the rectangular wave-guide. The cap and the bottom surface of the wave-guide form a localized radial cavity around the diode. The mm wave signal generated by the diode is coupled to the main rectangular wave-guide cavity through the vertical open edges of the radial cavity.

3.22 HEAT SINK OF IMPATT OSCILLATOR

IMPATT diode is a solid-state device of small dimension and operates at high current density. The conversion efficiency of the device is low so that a small fraction of the dc input power is converted into mm wave power and the rest is dissipated as heat raising the junction temperature of the device. Without adequate heat sink arrangement, the junction temperature increases the reverse saturated current, which leads to thermal runaway and destruction of the device. The heat sink in which the packaged diode is mounted must provide adequate heat flow path to the ambient from the diode mounted in a heat sink and maximum power P_{max}, which can be dissipated is given by

$$P_{max} = \frac{(T_j - T_0)}{\theta_T} \tag{3.14}$$

where T_j = junction temperature , T_0 = ambient temperature, and θ_T = total thermal impedance of the diode and the heat sink.

The thermal impedance is defined as the temperature rise at the junction by the dissipation of one watt of power. The size, shape and thermal conductivity of the diode and its metal heat sink determine the thermal impedance. Heat flows from the junction through metal contacts into a copper block and to the ambient through the surface of the copper heat sink. For Ka-Band operation the input power is less than 8 W, a heat sink of dimension 3.8 cm × 3.8 cm × 8 cm is adequate for safe operation.

3.23 BIAS FILTER OF THE IMPATT OSCILLATOR

The diode needs dc bias current at breakdown voltage from a dc power supply. This should be fed through a low pass filter as shown in Figure 3.25, which allows the dc bias current to pass through the diode but prevents the mm wave signal generated by the diode to pass through the power supply. The bias filter is constructed by two sections of $\lambda/4$ low impedance lines with inner conductors surfaces separated from the inner surfaces of the outer co-axial line by means of thin Teflon tape which then act as capacitors. In between the two low impedance sections, a high impedance line is formed using a narrow radius for inner conductor. The three section together comprise a low pass π type filter and the capacitance and the inductance values of the filter is given by

$$C = \left\{ \frac{0.556\varepsilon'}{\ln\left(\frac{b}{a}\right)} \right\} \tag{3.15a}$$

$$L = 2.0 \times 10^{-9} \ln\left(\frac{b}{a'}\right) \text{H/cm} \tag{3.15b}$$

where

a = Outer radius of inner conductor of the capacitive sections
b = Inner radius of the outer conductor
a' = Outer radius of the inner conductor of the middle section.

The cut-off frequency of the filter is given by

$$F_c = \frac{1}{\left[2\pi(LC)^{1/2}\right]} \tag{3.16}$$

where C is the equivalent capacitance.

3.24 OUTPUT CHARACTERISTICS OF A mm WAVE IMPATT SOURCE

The block diagram of the experimental set-up for studying the various characteristics of Ka-band IMPATT oscillators is shown in Figure 3.26. The output power and frequency of the

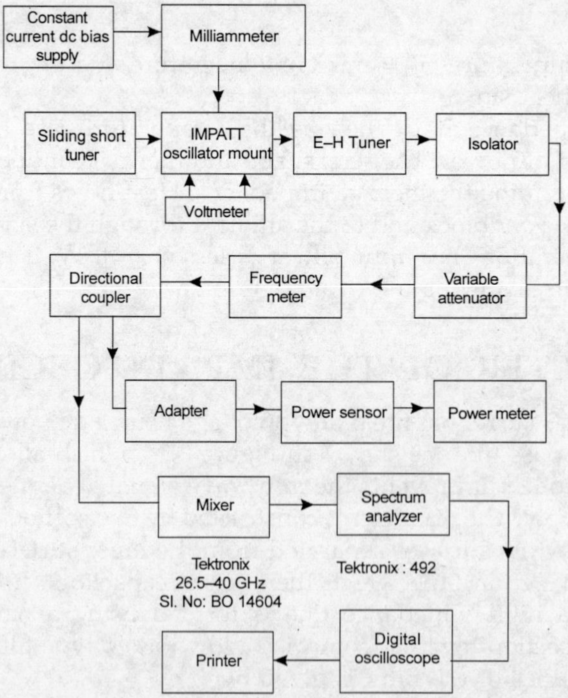

Figure 3.26 Experimental set up for studying the various characteristics of Ka-Band IMPATT oscillator.

oscillator can be controlled by adjustment of the cap height and cap diameter (cavity tuning) and by varying the dc current (electronic tuning) and also by varying the position of the sliding short tuner behind the mount (mechanical tuning). Typical results of measurement of power and frequency for various dc currents are shown in Figures 3.27 and 3.28 respectively for various values of cap diameter and cap height. Typical frequency spectrum of a Ka-band

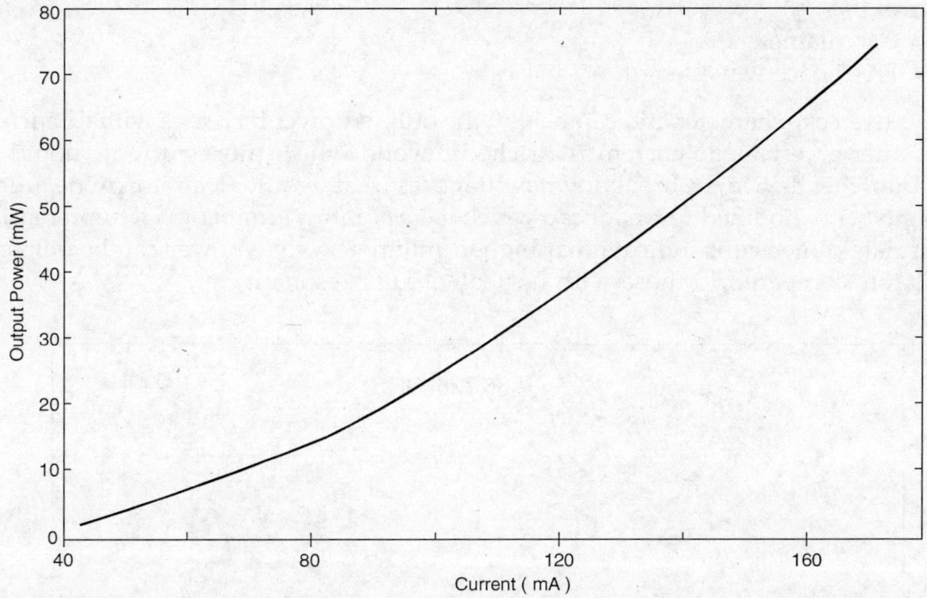

Figure 3.27 Current versus output power of Ka-band Si SDR IMPATT source.

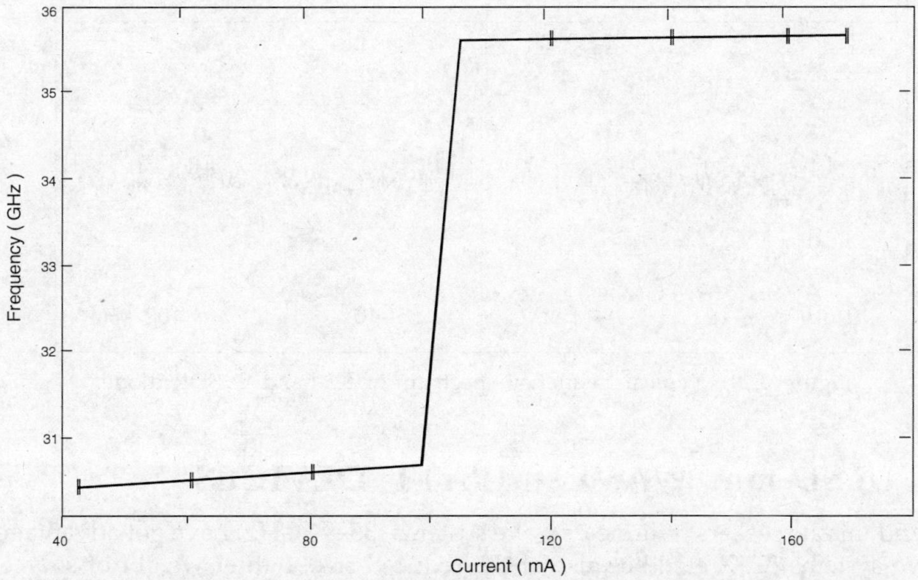

Figure 3.28 Current versus frequency of Ka-band Si SDR IMPATT source.

SDR diode is shown in Figure 3.29. The results indicate that the frequency increases with decreasing cap diameter. The relation between the frequency and the cap diameter can be calculated considering the field propagating in radial line and is given by

$$D = \frac{\lambda_0}{2} \tag{3.17}$$

where
D = Cap diameter
λ_0 = Free space wave length.

For a fixed cap diameter and cap height, the output power increases with dc current and saturates after a certain dc current is reached beyond which more current should not be pushed into the device at breakdown voltage as that would lead to diode burn out. Experiments with modified resonant cap which incorporate a number of circumferential slots of appropriate dimensions indicate that higher millimeter wave power can be delivered by the oscillators compared to those with normal circular resonant cap.

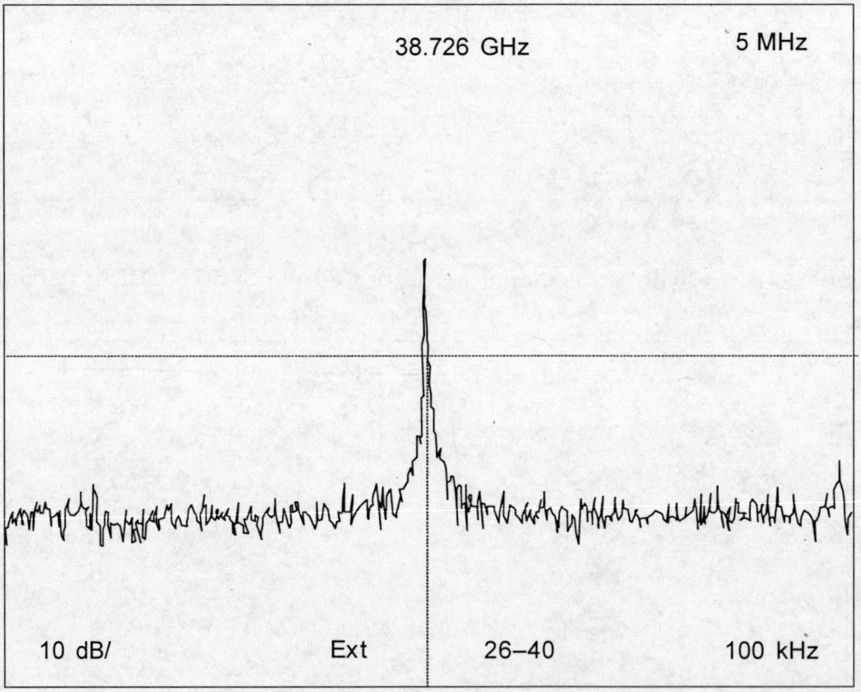

Figure 3.29 Typical frequency spectrum of Ka-band Si SDR diode.

3.25 PULSE mm WAVE IMPATT DEVICES

In radar and missile seekers millimeter wave systems (35–94 GHz) have got advantages over microwave systems (C, X etc.) because of the reduced size, lighter weight of sources with greater resolutions and accuracy accompanied by smaller antennas. Millimeter waves can penetrate through cloud, smoke and dust at window frequencies of 35 GHz (Ka-Band) and 94 GHz (W-Band), which indicate their suitability over electro optical or infra red systems as radar seekers. For use as radars of various kinds, pulsed mm wave sources are essential and solid state IMPATT pulsed sources have recently been developed with sufficient pulsed output power. Pulsed power output of 23 W at 35 GHz, 42 W at 94 GHz have been obtained from double drift flat profile diodes based on silicon.

3.26 FABRICATION OF PULSED IMPATT DIODES

Silicon multiple epitaxial technique is generally used to fabricate double drift pulsed diodes. Starting from an n^{++} substrate an n^- layer is just grown on it, followed by the growth of a p-layer using molecular beam epitaxy technique. This is followed by a shallow p^{++} boron diffusion or ion implantation. The silicon wafer is then metallized on both sides by Titanium/ Chromium followed by Palladium/Platinum and then by Gold. Circular mesa diodes of appropriate diameter (50 μm for 36 GHz, 20 μm for 94 GHz) are then obtained using photolithography and etching. Diode chip is then bonded to a gold plated copper disc using thermo compression bonding and ultrasonic vibration. The bonded chip is then packaged in a small ceramic ring in a wire-bonding machine by bonding of gold ribbons. Gold-coated brass or copper caps finally cap the package.

The heat generation in a pulsed IMPATT diode is much lower than CW IMPATT diodes carrying the same current due to the low duty cycle, and hence large amount of current may be allowed to pass through the device under pulsed biased condition.

From Conductance–Susceptance plots of an IMPATT diode it is seen that peak negative conductance increases with increase in dc current, but at high value of dc current it decreases due to the effect of space charge, which leads to the widening of the avalanche zone. The maximum current density is limited for CW diodes by the thermal heating of the device. But maximum current density can be increased for pulsed diodes to a much larger value until the space charge effect comes into play. Thus the pulse width and the duty factor as well as the device impedance would determine the optimum dc current density. The diode structure and the operating current can be determined by using the small signal or more accurately, by using the large signal analysis considering the space charge effect.

3.27 FREQUENCY CHIRP OF PULSED IMPATT OSCILLATORS

The frequency of IMPATT oscillators changes during the bias pulse due to the rise of temperature caused by heavy bias current pulses. For a flat current pulse, the diode junction is at lowest temperature at the beginning of the pulse, which heats up at the end of the pulse. The frequency change during the pulse is called the frequency chirp. The temperature variation depends mainly on current density and junction area and thermal resistance.

For a flat current pulse the oscillation frequency decreases during the pulse. By using an upward ramp on the flat current pulse usually reduces frequency chirp. By adjusting the slope of the current pulse, thermal and current effects would cancel and chirp can be much reduced. (Fig. 3.30).

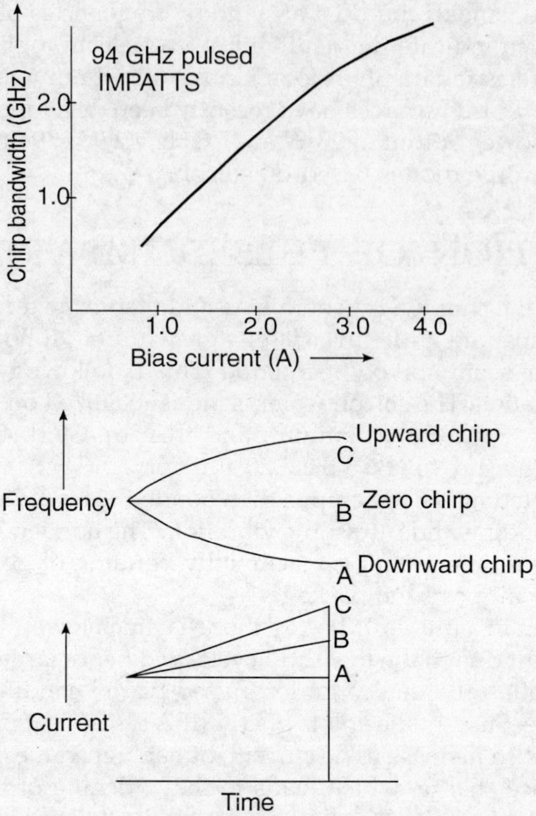

Figure 3.30 dc current ramping for controlling frequency chirp.

The pulse modulator of the pulsed IMPATT source should provide sharp rising current pulses up to 15 A with stability. There should be arrangement for proper adjustment regarding the control of the magnitude and slope of the current pulse to minimize chirping by current ramping.

3.28 DISCUSSIONS

Fundamental physical processes in the depletion layer of IMPATT diodes have been discussed in this Chapter. The origin of microwave oscillations in IMPATT diodes have been explained along with small signal properties and efficiency of the device. Various IMPATT doping profiles have been presented. Also the space charge effect, and heat sink in connection with IMPATT operation have been discussed. The various process steps for fabrication of IMPATT diodes and have also been presented in this Chapter. Various aspects of CW and pulsed mm wave IMPATT sources Discussions have been presented regarding doping profiles and basic properties at several window frequencies followed by the technology of fabrication through different process routes and packaging of the devices. A Ka-band mm wave oscillator has then been described for a compact resonant cap cum wave-guide heat sink mount with output characteristics. Pulsed IMPATT diodes have then been discussed with important characteristic features of frequency chirp and the method of controlling the same.

REFERENCES

1. Iglesius, D.E., Irvin, J.C. and Niehaus, W.C., 10 W and 12 W GaAs IMPATTs , *IEEE Trans. Electron Devices,* **ED-22:** 200, 1975.
2. Ishibashi, T., et al., Liquid Nitrogen cooled submillimeter wave silicon IMPATT diodes, *Electron. Lett.,* **13:** 299–300, 1977.
3. Johnston, R.B., Deloach, B.C. and Cohen, B.G., A silicon diode microwave oscillator, *Bell Syst. Tech. J.,* **44:** 369–372, 1965.
4. Read, W.T., A proposed high frequency negative resistance diode, *Bell Syst. Tech. J.,* **37:** 401–446, 1958.
5. Cannali, C., Ottaviani, G. and Quaranta, A. A., Drift velocity of electrons and holes and associated anistrophic effects in silicon, *J. Solid State Physics,* **32:** 1707, 1971.
6. Houston, P.A. and Evans, A. G. R., Electron drift velocity at n-GaAs at high electric field, *Solid State Electron,* **20:** 197–204, 1977.
7. Dalal ,V.B., Hole velocity in p-GaAs, *Appl. Physics, Lett,* **16:** 489–491, 1970.
8. Levinshtein, R.E., Recent results of studies of inter-valley transfer of hot electrons, *Sov. Physics Semicond.,* **13:** 735–745, 1979.
9. Cook, L.W., Bullman, G.E. and Stillman, G.E., Electron and hole Impact ionization coefficients in InP determined by photo multiplications measurements, *Appl. Physics Lett.,* **40:** 589–591, 1982.
10. Ito, M., et al.; Ionization rates for electrons and holes in GaAs, *J. Appl. Physics,* **49:** 4607–4608, 1978.
11. Grant, W. N., Electron and hole ionization rates in epitaxial silicon at high electric field, *Solid State Electron,* **16:** 1189–1203, 1975.
12. Logan, R.A. and Sze, S.M., Avalanche multiplication in Ge and GaAs p-n junctions, *J. Physics Soc. (Japan) Suppl.,* **21:** 434, 1966.
13. Sze, S.M. and Gibbons, G., Avalanche breakdown voltage of abrupt and linearly graded p–n junction in Ge, Si, GaAs and GaP, *Appl. Physics Lett.,* **8:** 111, 1986.
14. Datta, D.N., et al, Computer analysis of dc field and current density profiles of DAR IMPATT diode, *IEEE Trans Electron Devices,* **ED-29:** 1813–1816,1982.
15. Gilden, M. and Hines, M.E., Electron tuning effects in the Read microwave avalanche diodes, *IEEE Trans. Electron Devices,* **ED-13:** 169–175, 1966.
16. Berenz, J.J., Frank, F.B. and Heirel, T.T., Ion implanted p-n junction indium phosphide IMPATT, *Electron Lett.,* **14:** 683–684, 1978.

17. Midford, T.K. and Bernick, R.L., Millimeter wave IMPATT diodes and oscillators, *IEEE Transactions on Microwave Theory and Techniques*, **MTT-27**, 1979.

18. Schrafetter, D.L and Gummel, H.K., Large signal analysis of a silicon Read diode oscillator, *IEEE Trans Electron Devices,* **ED-18:** 64–77, 1969.

19. Sridharan, M. and Roy, S.K., Computer studies on the widening of the avalanche zone and decrease of efficiency in silicon *X*-band sym. DDR, *Electron Lett.*, **14:** 635–637, 1978.

20. Sridharan, M. and Roy, S.K., Effect of mobile space charge on the small signal admittance of silicon DDR, *Solid State Electron,* **23:** 1001–1003, 1980.

21. Majumder, N. and Roy, S.K., Studies on pulsed mm wave low-high-low silicon IMPATT diodes at high current density and dependence on structural parameters, *Semiconductor Sci and Technology*, **12:** 623–630, 1997.

22. Elta, M.E. and Hadded, G.I., High frequency limitations of IMPATT, MITAT and TUNNETT mode devices, *IEEE Trans. Micro, Theory Tech.*, **MTT-27:** 442–449, 1979.

23. Mitra M,. et al., Experimental studies on the process steps for fabrication of IMPATT diodes and corresponding study of dc breakdown voltage, *IEEE Tech. Rev.* (*India*) **10 (4):** 351–354, 1993.

24. Cowley, A.M., et al., *IEEE, Journal of Solid State Circuits,* **Sc 5:** Dec. 1970, pp. 338–345.

25. Alderstein, M.G., Wallace, R.N. and Steel, S.R., Millimeter wave GaAs Read IMPATT diodes, *IEEE Trans. Electron Devices,* **ED-25:** 1151–1156, 1978.

26. Vasudev, P.K., Recent advances in device processing and packaging of high power pulsed GaAs double drift IMPATTs at X-band, *IEEE Trans. Electron Devices,* **ED-31:** 1044–1048, 1984.

27. Chang, K. et al., GaAs Read type IMPATT diode for 130 GHz CW operation, *Electron Lett.,* **17:** 471–472, 1981.

28. Mazumder., N., and Roy, S.K., Studies on pulsed mm wave silicon IMPATT diodes at high current density and dependence on structural parameters, *Semiconductor Science and Technology*, **12**, 623–630, 1997

SUGGESTED READINGS

1. Sze, S.M., *Semiconductor Devices: Physics and Technology*, 2nd ed., John Wiley & Sons, 2002.

2. Liao, S.Y., *Semiconductor Electronic Devices*, Prentice-Hall Inc., Englewood Cliffs, N.J., 1990.

3. Liao, S.Y., *Microwave Devices and Circuits*, 3rd ed., Prentice-Hall of India, 1995.

4. Roy., S.K., Transit time devices, Wiley Encyclopedia of Electronics and Electrical Engineering, Ed., John Webbler, Vol 22, pp. 421–442, May, 1999.

5. Fong., T.T. and Kumo., H.J., Millimeter wave IMPATT sources, *IEEE, Transaction on Microwave Theory and Techniques*, **MTT-27**, May, 1979.

PROBLEMS

1. Draw the schematic diagram of an IMPATT diode and explain the two effects that combine to produce 180 degrees of phase difference between the applied voltage and the resulting current pulse. Describe one wave guide mounted IMPATT diode microwave oscillator.

2. Mention some of the merits and demerits of an IMPATT diode.

3. An IMPATT diode has the following parameters, carrier drift velocity $v_d = 1.5 \times 10^7$ cm/sec, Drift region length L = 8 μm, maximum operating current I_{max} = 180 mA. maximum operating voltage V_{max} = 90 V, efficiency η = 10%. Calculate (a) the maximum CW output power in watts, (b) the frequency of the output power in GHz.

4. What are the major drawbacks of avalanche devices? What limitations does this place have on their applications?

5. Make a comparative study of Single Drift Region (SDR), Double Drift Region (DDR), and Double Avalanche Region (DAR) IMPATT diodes.

6. Explain why it is possible to obtain amplification by using a negative resistance device.

7. An IMPATT diode has a pulsed operating voltage of 150 V and a pulsed operating current of 0.5 A. The efficiency is about 10%. Calculate (a) the output power, (b) the frequency of oscillation produced, if the duty cycle is 15% and pulsed width is 0.02 ns.

8. Draw the resonant cap oscillator arrangement of an IMPATT diode and explain its operation.

9. Draw the circuit diagram of IMPATT diode amplifier and explain its operation.

10. Mention some specific applications of IMPATT diode.

11. Mention some of the advantages of millimeter wave propagation over microwave.

12. Mention some of the constructional difference between mm wave IMPATT and microwave IMPATT.

13. Write down the process steps for fabrication of a mm wave IMPATT.

14. Mention some of the advantages and applications of pulsed mm wave IMPATT diode.

15. What is frequency chirp and how can it be reduced?

Chapter 4

TRAPATT Diodes

4.0 INTRODUCTION

TRAPATT is a solid state microwave device having a p^+nn^+ or n^+pp^+ structure similar to IMPATT which produces high microwave power with very high dc to microwave conversion efficiency (40–60%). TRAPATT is the acronym for Trapped Plasma Avalanche Triggered Transit. The device is capable of producing very high power in pulsed operation. Although TRAPATT is somewhat similar to IMPATT in structure, its mode of operation is distinctly different from that of IMPATT and it exhibits much higher efficiency than IMPATT diodes. TRAPATT operation with high efficiency of dc to microwave conversion was first reported in 1967 from RCA Laboratories by Prager, Chang and Weisbrad [1]. The TRAPATT mode of oscillation is characterized by a lower oscillation frequency than the IMPATT mode for which the frequency is determined by the transit time of charge carriers to cross the depletion layer with saturated drift velocity.

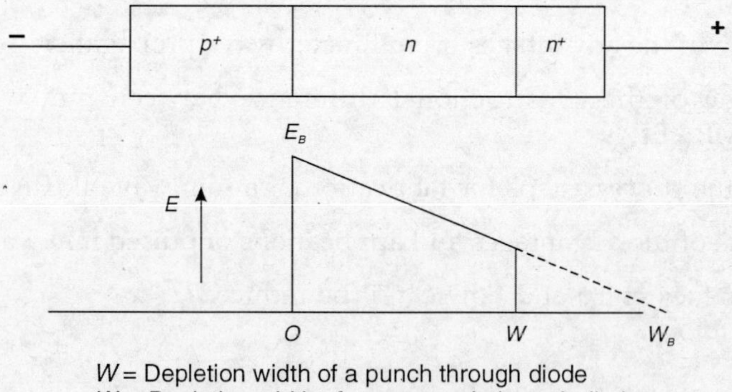

W = Depletion width of a punch through diode
W_B = Depletion width of a non punch through diode
F = W_B/W = punch through factor

Figure 4.1 Electric field profile of a reverse biased flat profile $p^+n\ n^+$ diode.

After the discovery of TRAPATT oscillation, the physics of its operation was explained a year later through computer simulation by Johnson, Scharfetter and Bartelink [2]. TRAPATT diode needs considerable punch through under reverse bias condition (Figure 4.1) for excitation of TRAPATT mode. The term punch through means that the electric field under reverse bias condition is quite high throughout the depletion layer, which becomes abruptly zero at the epitaxy substrate interface (nn^+ or pp^+) unlike the usual IMPATT structure. The field is highest at p^+n junction in case of p^+nn^+ diode and it linearly falls up to nn^+ interface where it abruptly becomes zero.

4.1 BASIC PRINCIPLE OF TRAPATT OPERATION

Computer simulation indicates that a trigger voltage applied across the diode leads to an avalanche formation at the high field point (p^+n junction plane) which then propagates as a transient avalanche zone to the other end nn^+ of the epitaxial n-layer.

Electron hole plasma of very high density is created by the avalanche electric field initially at the junction end, which causes the field to collapse due to sharp increase of conductivity in that region. As a result the electrons and holes can drift very slowly and hence the plasma is trapped.

The avalanche zone travels faster across the depletion layer like a shock front and fills the entire depletion layer with dense plasma of electrons and holes. The electrons and holes in the depletion layer then begin to drift in opposite direction with low drift velocity due to the low field in the plasma, electrons moving towards the positive side of the reverse biased device. With extraction of the plasma the electric field in the drift layer rises and the plasma is extracted more rapidly. After total extraction of the plasma the field recovery is complete and the original field profile is re-established and another avalanche shock front then passes through the depletion layer. Thus the cycle repeats leading to TRAPATT oscillations comprising of a high field state with zero current followed by a low field state with heavy current after formation of plasma.

4.2 AVALANCHE SHOCK FRONT PROPAGATION

To understand the propagation of the avalanche shock front through the depletion layer, let us consider the field profile of a reverse biased p^+nn^+ diode biased below the threshold of breakdown with epitaxial n-layer having uniform or flat doping shown in Figure 4.1.

When a voltage greater than the dc breakdown voltage is rapidly applied across the terminals, a large displacement current I passes through the diode. The depletion layer does not contain any mobile carrier and hence, there is only ac displacement current which is given by

$$I = \varepsilon \frac{\Delta E}{\Delta t} \tag{4.1}$$

where ε is the dielectric constant of the semiconductor.

The field profile therefore rises till it exceeds the static breakdown value at time t_2 as shown in Figure 4.2. The time variation of the field profile is given by Equation 4.1.

The avalanche process and electron hole plasma have not started yet due to sudden application of the trigger voltage. The corresponding electric field profile at time t_2 is shown in Figure 4.2.

Note that the slope of the field remains the same, which is determined by the doping density in the n-epitaxial layer, and there is a punch through in the nn^+ interface where the field abruptly becomes zero in the n^+ substrate.

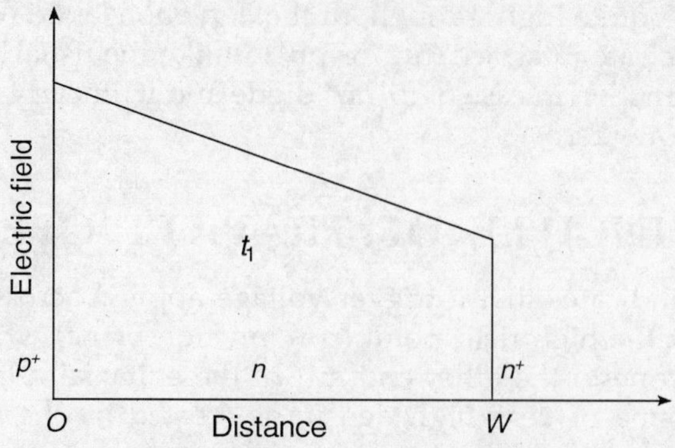

Figure 4.2 Field profile in the depletion layer at t_1 before the onset of the plasma.

At a later time t_3, the avalanche has started and a dense plasma of electrons and holes is created which leads to the collapse of the field to nearly zero due to the high conductivity of plasma as shown in Figure 4.3. In front of the plasma, the avalanche has not started yet and the field continues to rise up due to the displacement current in the carrier free portion of the depletion layer.

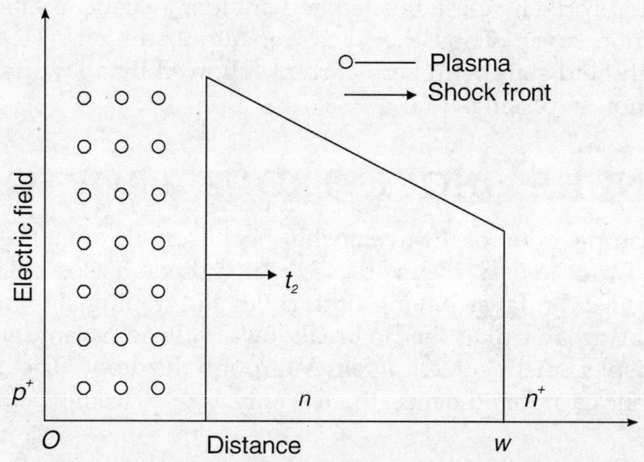

Figure 4.3 Field profile at t_2 when the shock front is propagating in this depletion layer leaving behind a dense plasma.

In the plasma region the current is due to the conduction current, arising from the movement of electrons and holes in opposite directions with low field mobility. The avalanche zone spreads further to the right at a later time t_4 with the generation of e-h plasma and the field continues to rise in the carrier free drift region as shown in Figure 4.4. Thus the avalanche shock front reaches the substrate with a high velocity and fills the whole of the depletion layer with dense plasma of electrons and holes which gets almost trapped due to the collapse of the field.

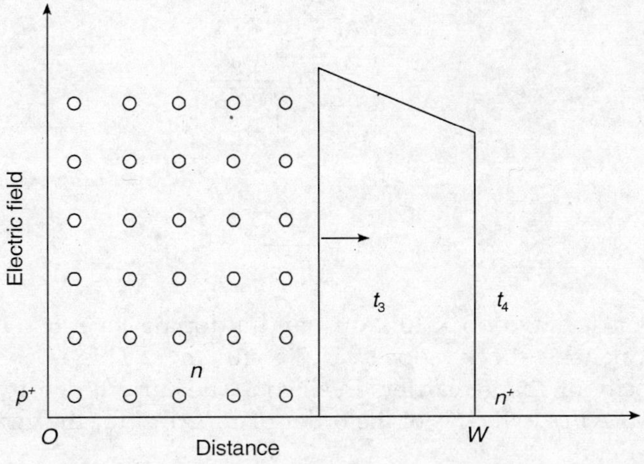

Figure 4.4 Field profile at a a later time when the shock front is reaching the substrate.

4.3 VELOCITY OF THE AVALANCHE SHOCK FRONT

We would now derive an expression for the avalanche shock front velocity. The electric field in the carrier free drift region linearly falls in the depletion region if the doping N_D is constant. The slope can be obtained from Poisson's equation:

$$\nabla D = qN_D \tag{4.2}$$

where

D = Electronic displacement
N_D = Doping density in the n-epitaxial layer
q = Electric charge

In one dimensional case

$$\varepsilon \frac{\Delta E}{\Delta x} = qN_D$$

or

$$\frac{\Delta E}{\Delta x} = \frac{qN_D}{\varepsilon} \tag{4.3}$$

The field rises in the carrier free drift region during shock front propagation

$$\frac{\Delta E}{\Delta t} = \frac{I}{\varepsilon} \tag{4.4}$$

The shock front location is given by the point at which the field curve rises to breakdown field value E_B with the slope given by Equation (4.3).

Thus the shock front velocity $v_z = \dfrac{\Delta E}{\Delta t}$ is given by

$$\frac{\Delta x}{\Delta t} = \frac{\left(\dfrac{\Delta E}{\Delta t}\right)}{\left(\dfrac{\Delta E}{\Delta x}\right)} = \frac{\left(\dfrac{I}{\varepsilon}\right)}{\left(\dfrac{qN_D}{\varepsilon}\right)}$$

or

$$v_x = \frac{I}{qN_D} \tag{4.5}$$

For a typical current density of 3×10^8 Am^{-2} and a doping level 6×10^{21} m^{-3} of 2–4 GHz TRAPATTs, the avalanche shock velocity is 3×10^5 m/s. This is about three times the saturated drift velocity of charge carriers in silicon. The current density is much larger for TRAPATTs than IMPATTs, which is of the order of 10^6 Am^{-2} for the generation of the same frequency of oscillation.

4.4 EXTRACTIONS OF TRAPPED PLASMA AND RECOVERY OF THE ELECTRIC FIELD

After the passage of the avalanche shock front, the generated electrons and holes in the plasma begin to drift apart in opposite directions with low plasma velocity determined by low field mobility. For the p^+nn^+ TRAPATT structure, electron extraction first starts at the p^+ terminal and the hole extraction starts a slightly later time (w/v_z) at the n^+ terminal when the shock front has arrived at the substrate. The plot of the current against time through the device is shown in Figure 4.5, which is assumed to be a constant rectangular wave form equal to that in avalanche transient period.

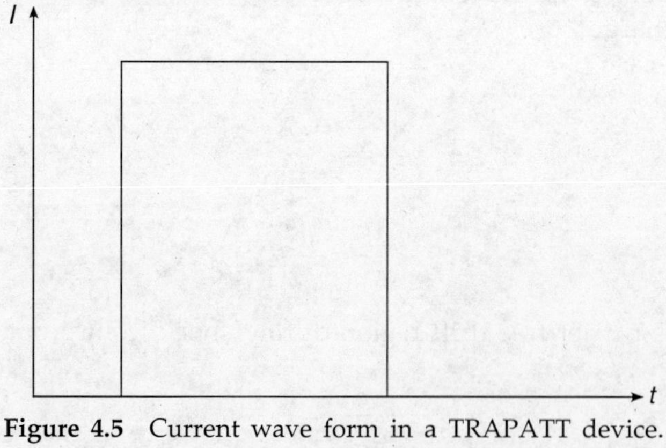

Figure 4.5 Current wave form in a TRAPATT device.

The corresponding voltage wave form depends on the changing field profile as the plasma is extracted from the active region and is shown in Figure 4.6. The field collapses to a low value everywhere in the depletion layer after the avalanche shock front has passed. This is shown in Figure 4.7, where the high density of electrons and holes are accompanied by low field. The low field present in the plasma drives electrons towards the p^+ terminal and

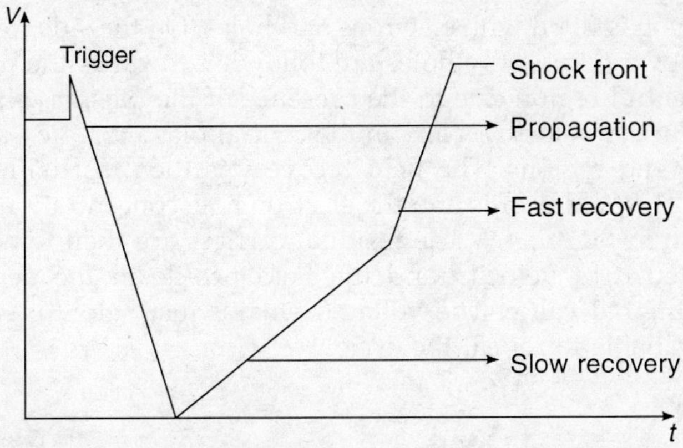

Figure 4.6 Voltage wave form for the TRAPATT device.

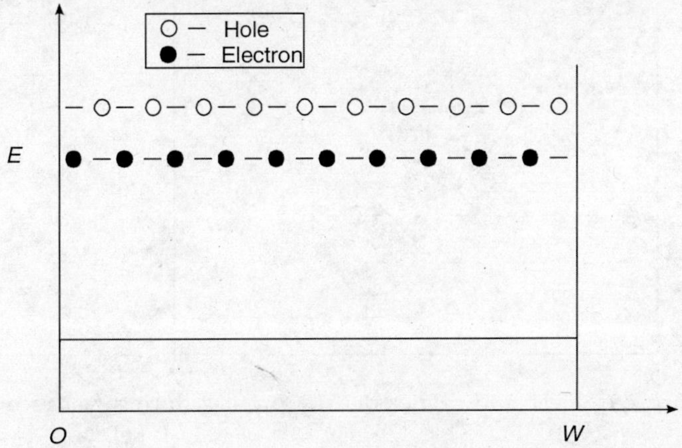

Figure 4.7 Field profile electron and hole density profiles in the plasma state.

the holes towards the n^+ terminal and the charge extraction causes the field to rise. The field profile and the carrier concentration in the early stages of slow charge recovery is shown in Figure 4.8.

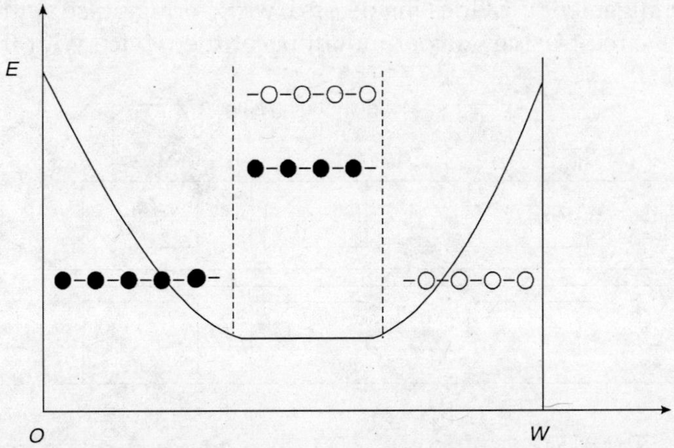

Figure 4.8 Field and current profiles during slow recovery.

The central region is filled with electrons and holes. On the side towards the p^+ junction only electrons are flowing and only holes are following towards the n^+ substrate. Thus the field is low in the central region due to the presence of the plasma whereas it linearly rises on either side. The carrier velocity is low in the central plasma region and rises to saturated drift velocity in the end regions. The field recovery is then fast with rapid extractions of electrons and holes. The field profile and the electron hole concentrations in the second phase of recovery is shown in Figure 4.9. The residual carriers are then removed from the active region with saturated drift velocity and the field profile in the depletion layer is then established to the original value. The voltage remains just below breakdown till the next trigger voltage is available to repeat the cycle.

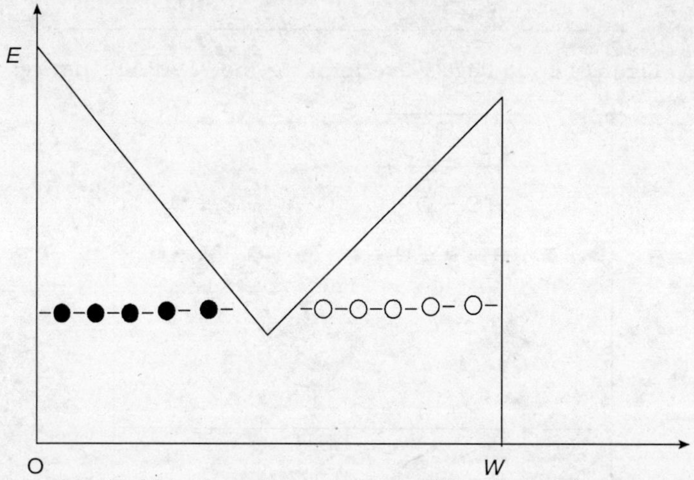

Figure 4.9 Field and carrier density profiles during fast recovery.

4.5 TRIGGERING VOLTAGE AND TRAPATT CIRCUIT

At the beginning of each cycle of TRAPATT oscillation, the device needs a triggering pulse, which provides the necessary voltage to initiate an avalanche shock front. The TRAPATT oscillator circuit arrangement provides the necessary triggering pulse at the beginning of each cycle. A TRAPATT circuit using coaxial mounting of the device with three sling tuners is shown in Figure 4.10.

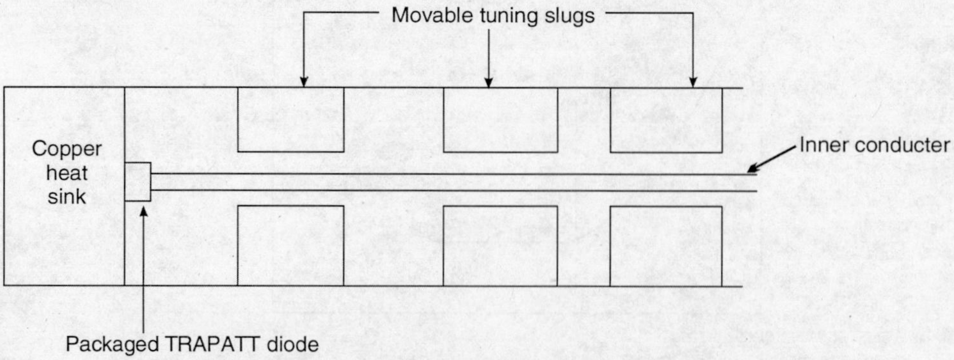

Figure 4.10 Coaxial circuit TRAPATT oscillator with tuning slugs.

The slide tuner nearest to the device is positioned at a distance of $\lambda/2$ from the device where λ is the wavelength at TRAPATT oscillation frequency. The circuit can also produce oscillation at IMPATT mode which is, however, quite a different mode than TRAPATT. IMPATT oscillation starts from small signal amplitudes and builds up to large signal whereas TRAPATT oscillation is a large signal oscillation produced by triggering pulses. In the TRAPATT circuit shown in Figure 4.10, the triggering pulses are generated in the following way. Initially with increase of bias current IMPATT oscillations are generated by the device placed in the cavity which grows into rf voltage of appreciable magnitude and provides the first triggering pulse. It produces an avalanche shock front and the first trapped plasma state for the TRAPATT mode of oscillation. The voltage across the device abruptly drops, which is placed at the end of the transmission line. The negative voltage pulse travels towards the first low impedance tuning slug and is reflected at that point. The pulse changes sign on reflection and become a positive pulse travelling towards the device. If the first slug is placed at a distance $\lambda/2$ from the diode, the reflected positive pulse arrives at the device to produce the triggering pulse of necessary voltage at a full time period later (caused by the travel of unit wavelength along the transmission line), to produce the next avalanche shock front and the trapped plasma state. This again produces a sharp rectangular voltage pulse, which similarly travels down the transmission line and comes back after reflection to produce the next positive triggering pulse. Thus the triggering pulses are continuously produced resulting in TRAPATT oscillations in the device.

4.6 TRAPATT OSCILLATION CONDITION AND OSCILLATION FREQUENCY

For TRAPATT operation it is required that the field should rise to the breakdown field E_B or slightly less than it at the end of that recovery period. It can be shown that to avoid the premature avalanching, i.e., rise of the electric field to slightly less than about E_B at the end of charge extraction is necessary for which a punch through p^+nn^+ structure for the TRAPATT device is required. In a punch through structure the electric field profile at reverse breakdown does not linearly reach zero value at substrate end of the depletion layer but abruptly becomes zero at the nn^+ interface as shown in Figure 4.1.

The avalanche shock front passes through the depletion layer in a short transit time τ_A given by $\tau_A = w/v_z$ where w = depletion layer width, and v_z = shock front velocity. An approximate expression of the charge extraction time, corresponding to the slow and fast recovery periods can be obtained in the following way. Assume that the low field mobility μ_p of electrons and holes in the plasma created by the passage of avalanche shock front to be identical. The corresponding drift velocity of both the carriers then become $v_p = \mu E_p$, when E_p is low field in the e-h plasma. During slow recovery the carriers travel $w/2$ on either side after which only electrons are present on the left side of the central phase and holes on the right side.

The slow recovery period is given by

$$T_{\text{slow}} = \frac{w}{2v_p} \tag{4.6}$$

After slow recovery we assume that charge carriers quickly reach the terminals with saturated drift velocity v_s; the fast recovery period is thus given by

$$T_{fast} = \frac{w}{2v_s} \tag{4.7}$$

Thus the total charge extraction time T_E is given by

$$T_E = T_{slow} + T_{fast} = \frac{w}{2}\left(\frac{1}{v_p} + \frac{1}{v_s}\right) \tag{4.8}$$

During recovery the depletion layer consist of two regions, central plasma region containing two types of carriers and side recovery regions containing one type of carrier. Poisson's equation in the recovery region containing electrons is given by

$$\frac{dE}{dx} = \frac{-q}{\varepsilon}(n_R + N_A) \tag{4.9}$$

where

n_R = Mobile electron density in the recovery region
N_A = Immobile impurity charge density.

Consider a frame of reference moving with the velocity of the point of intersection of the electron recovery region and the plasma region; the electron flow in the recovery region can be equated to the electron flow in the plasma region. Thus

$$n_R(v_p + v_s) = n_p(v_p + v_p) = 2n_p v_p$$

Thus

$$n_R = \frac{2n_p v_p}{v_p + v_s} \tag{4.10}$$

The field E in the recovery region rises with time according to

$$\frac{dE}{dT} = \frac{v_p dE}{dx} \tag{4.11}$$

since the intersection point is moving with velocity v_p.

Thus after the extraction time, the maximum field rises to

$$E = E_p + \frac{qv_p T_E}{\varepsilon}[n_R + N_A] \tag{4.12}$$

An expression for n_R can be derived in the following way. The hole density in the plasma differ slightly from electron density by ionized impurity density to maintain the charge neutrality in the plasma.

Thus

$$p_p = N_A + n_p \tag{4.13}$$

Consider now an avalanche shock front proceeding in the depletion region leaving behind electron hole plasma. Current in the plasma region is given by

$$\begin{aligned} I &= qv_p\,(n_p + p_p) \\ &= qv_p\,(2p_p + N_A) \\ &= qv_p\,(2n_p - N_A) \end{aligned} \tag{4.14}$$

This is equal in magnitude to the displacement current in front of the shock front

$$I = qN_A v_z \tag{4.15}$$

Thus

$$\begin{aligned} qN_A v_z &= qv_p(2p_p + N_A) \\ &= qv_p(2n_p - N_A) \end{aligned} \tag{4.16}$$

Hence
$$p_p = \frac{N_A}{2\left(\dfrac{v_z}{v_p} + 1\right)} \tag{4.17}$$

and
$$n_p = \frac{N_A}{2\left(\dfrac{v_z}{v_p} - 1\right)} \tag{4.18}$$

Thus

$$n_R = \frac{v_p N_A}{v_p + v_s}\left(\frac{v_z}{v_p} - 1\right) \tag{4.19}$$

Hence substituting the values of n_R and T_E we have the following value of E at the end of the recovery period

$$E = E_p + \left[\frac{qv_p.w}{2\varepsilon\left(\dfrac{1}{v_p} + \dfrac{1}{v_s}\right)}\right]\left[\left(\frac{v_p N_A}{v_p + v_s}\right)\left(\frac{v_z}{v_p - 1}\right) + N_A\right] \tag{4.20}$$

This must not rise above the breakdown field E_B

Thus

$$E_B - E_p \geq \left[\frac{qwv_p\left(\dfrac{1}{v_p} + \dfrac{1}{v_s}\right)}{2\varepsilon}\right]\left[\left(\frac{v_p N_A}{v_p + v_s}\right)\left(\frac{v_z}{v_p - 1}\right) + N_A\right] \tag{4.21}$$

Since

$$v_s \gg v_p \text{ and } \qquad E_p \ll E_B$$

we have

$$E_B \geq \left[\frac{qwv_p}{2\varepsilon v_s}\right]\left[\frac{N_A v_z}{v_s + N_A}\right]$$

or

$$\frac{2\varepsilon E_B}{qwN_A} \geq \left(\frac{v_z}{v_s} + 1\right) \tag{4.22}$$

Since

$$\frac{\varepsilon N_B}{qN_A} = w_B$$

= The depletion layer width of a non-punch through device.

Hence

$$2\frac{w_B}{w} \geq \frac{v_z}{v_s} + 1 \tag{4.23}$$

We have seen that the avalanche shock front velocity is greater than the saturated drift velocity of carriers.

Thus

$$2F \geq \frac{v_z}{v_s} + 1 \tag{4.24}$$

where F = Punch through factor = w_z.

Hence a punch through device is necessary for oscillation in TRAPATT mode.

For a typical TRAPATT

since $v_z = 3v_s$ and $F \approx 2$

The punch through factor can be expressed in terms of the breakdown voltage V_B of the non-punch through diode of same epitaxial doping density and the reverse bias voltage V_p required to fully deplete the depletion layer.

Thus

$$V_p = \frac{qN_A w^2}{2\varepsilon}$$

and

$$F = (V_B / V_P)^{1/2}$$

The punch through factor is important in designing the TRAPATT device. Thus the current in the trapped plasma state is given by

$$J \leq qN_A v_s (2F - 1)$$

Let us consider a square wave current having half period for low voltage high current trapped plasma state.

The dc current $J_{dc} = J/2$

Thus

$$J_{dc} \leq \frac{qN_A v_s}{2}(2F - 1)$$

This gives the upper limit of dc current for sustained oscillation. The condition that the shock front velocity should be greater than the saturated drift velocity gives the lower limit of dc current.

$$J_{dc} \geq \frac{1}{2} qN_A v_s$$

4.7 FREQUENCY OF OSCILLATION OF A TRAPATT DIODE

We note that the sum of the shock front period and the recovery period is approximately half the total time period of oscillation. Thus the frequency of oscillation for TRAPATT operation is given by

$$f = \frac{1}{2T}$$

where

$$\frac{T}{2} = \frac{w}{v_z} + \frac{w}{2}\left(\frac{1}{v_p} + \frac{1}{v_s}\right) = \frac{1}{2F}$$

The low value of plasma velocity $v_p = \mu E_p$ limits the operation of TRAPATT to the lower microwave range. Cottam [5] has calculated the optimum frequency of p^+nn^+ silicon TRAPATT to vary from 6 GHz to 1 GHz for depletion width varying from 2 to 6 μm and doping density vary from $6 \times 10^{15} - 1.5 \times 10^{15}$ cm^{-3}.

4.8 PERFORMANCE OF TRAPATT OSCILLATORS

In practice most of the experimental reports on TRAPATT oscillators are in frequency range of 0.5 to 10 GHz. This 1.2 kW output in the pulse mode has been obtained at 1.1 GHz by Liu and Risko [6] with an efficiency of 60%. Peak passes of 100 to 50 W have been obtained in the frequency range of 4–8 GHz with efficiency varing from 25–40%. Higher efficiency of 75% have been observed at 0.6 GHz [7].

REFERENCES

1. Prager, H.S., Chang, K.K., and Weisbrad, S., High power high efficiency silicon avalanche diodes at ultra high frequencies, *Proc. IEEE*, **55**, 586–587, 1967.
2. Johnson, R.L., Scharfetter, D.L., and Bartelink, D.J., High efficiency sub transit time oscillations in germanium in avalanche diodes, *Proc. IEEE*, **56**, 1611–1613, 1968.
3. Bartelink, D.J. and Scharfetter, D.L., Avalanche shock fronts in p-n junctions, *Appl. Physics Letter*, **14**, 320–323, 1969.
4. Chlorofeine, A.S., Ikola, R.J., and Napoli, N.S., A theory for the high efficiency mode of oscillation in avalanche diodes, *RCA Rev.*, **3** 397–421, 1969.
5. Cottam, J., Theory for high efficiency oscillations silicon avalanche diodes, *Electronics Letters*, **6**, 203, 1970.
6. Liu, S.G. and Risko, J. J., Fabrication and performance of kilowatt L-band avalanche diode, *RCA Rev.*, **31**, 3–19, 1970.
7. Kostichack, D.F., UHF avalanche diode oscillation providing 100 watts peak power end 75 per cent efficiency, *Proc. LEEE Letters*, **58**, 1282–1283, 1970.

SUGGESTED READINGS

1. Liao, S.Y., *Microwave Devices and Circuits*, 3rd ed., Prentice-Hall of India, New Delhi, 1994.

2. Weisglass, P., Avalanche and barrier injection devices in M.J.Howes and D.V. Morgan (Eds.) *Microwave Devices*, John Wiley & sons, London, 1976 (Chapter 3).

3. Sze, S.M., *Physics of Semiconductor Devices*, John Wiley & Sons, New York 1981.

4. Gibbons, G., *Avalanche Diode Microwave Oscillates,* Clarendon Press, Oxford, 1973.

5. Roy, S.K., *Transit Time Devices,* Wiley Encyclopedia of Electrical and Electronic Engineering (J.G.Webster, Ed.), **22**, 421–442, 1999.

PROBLEMS

1. Explain the propagation of avalanche shock front in the depletion layer of a TRAPATT diode with necessary diagrams.

2. In a TRAPATT diode the doping density is 5×10^{15} cm^{-3} and current density is 3×10^{14} Am^{-2}, find the value of the shock front velocity.

3. Explain slow recovery and fast recovery of the electric field in a TRAPATT diode with necessary diagram and find an expression for the total extraction time of the plasma.

4. How the necessary triggering pulses are generated for a TRAPATT oscillator. Explain the same with a diagram of co-axial TRAPATT circuit.

5. Show by necessary derivation that the TRAPATT must have punch through structure for oscillation to occur for TRAPATT mode.

6. For a TRAPATT oscillating with frequency of 2 GHz, find the approximate value of plasma velocity if the width of the depletion layer is 5 microns.

Chapter 5

BARITT Diodes

5.0 INTRODUCTION

BARITT is a transit time microwave device that generates low noise microwave power at lower microwave frequencies. It is the acronym for Barrier Injection Transit Time. It is normally a p^+np^+ or metal-n-metal (mnm) structure and is equivalent to a pair of diodes connected back to back, one of which is forward biased and the other is reverse biased. Microwave oscillation in mnm BARITT diode was reported by Coleman and Sze [1] and in p^+np^+ BARITT diodes by Chu and Sze [2]. Unlike IMPATT diodes charge carriers in transit of these devices are not generated by avalanche multiplication but by forward injection through the forward biased junction. Thus BARITT diodes produce power with much less noise and are useful in microwave receivers as local oscillator because of their low noise figures. The carriers in these devices have a low unsaturated drift velocity in a considerable portion of the drift region unlike IMPATT diodes. This causes carriers in BARITTs to have a large transit time leading to oscillation only at low microwave frequencies. It produces low-noise microwave power but its power output is restricted to lower microwave frequencies up to X-band.

5.1 BARITT STRUCTURE AND DC FIELD PROFILES

Two BARITT structures commonly used are p^+np^+ with two semiconductor junctions and mnm (metal-semiconductor(n)-metal) with two Schottky junctions as shown in Figures 5.1(a) and 5.1(b). In both the cases, one of the junctions is forward biased and the other is reverse biased when a voltage is applied across the terminals. When a small voltage is applied to

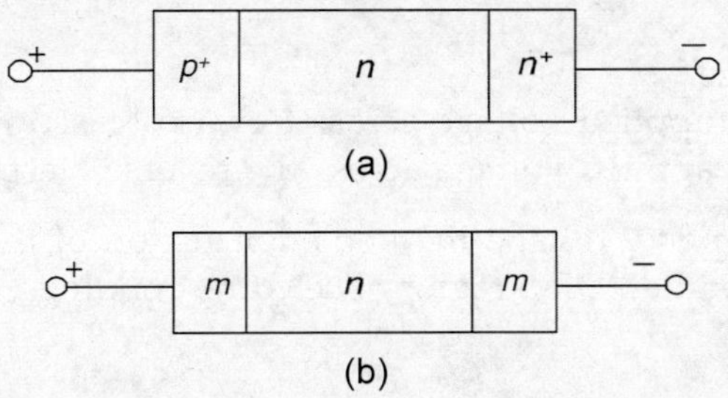

Figure 5.1 (a) p^+np^+ BARITT and (b) mnm BARITT.

either structure a small built in voltage develops across the forward biased junction A and the rest of the voltage appears across the reverse biased junction B with field profiles as shown in Figure 5.2.

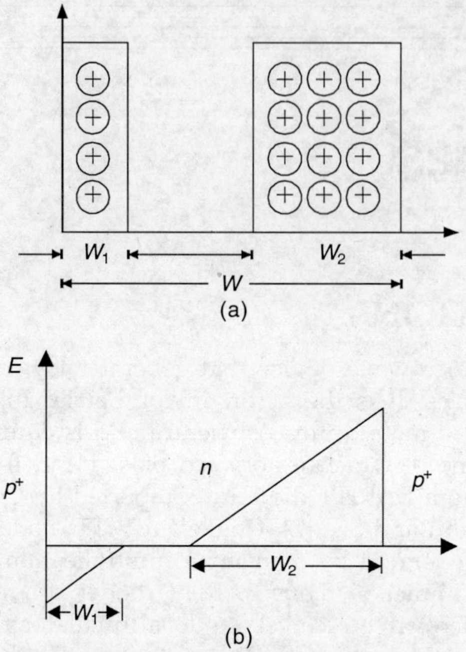

Figure 5.2 (a) Low-bias charge distribution and (b) field profile in a p^+np^+ for a low applied bias.

The width of the junction is given by

$$w_1 = \frac{2\varepsilon_s}{qN_D(V_{bi} - V_1)} \tag{5.1}$$

for the forward biased junction

and

$$w_2 = \frac{2\varepsilon_s}{qN_D(V_{bi} - V_2)} \tag{5.2}$$

for the reverse biased junction

where

V_{bi} = Built in voltage
V_1 = Portion of the applied voltage across forward biased junction
V_2 = Portion of the applied voltage across the reverse biased junction.

With increase in applied voltage the depletion region of the reverse biased junction reaches the forward biased depletion region. The corresponding voltage is known as punch through voltage.

At reach through

$$w_1 + w_2 = w$$

and

$$V_{RT} = \frac{qN_D}{2\varepsilon_s w^2 - w\left(\frac{qN_D V_{bi}}{\varepsilon_s}\right)^{1/2}} \tag{5.3}$$

The field profile at reach through is shown in Figure 5.3.

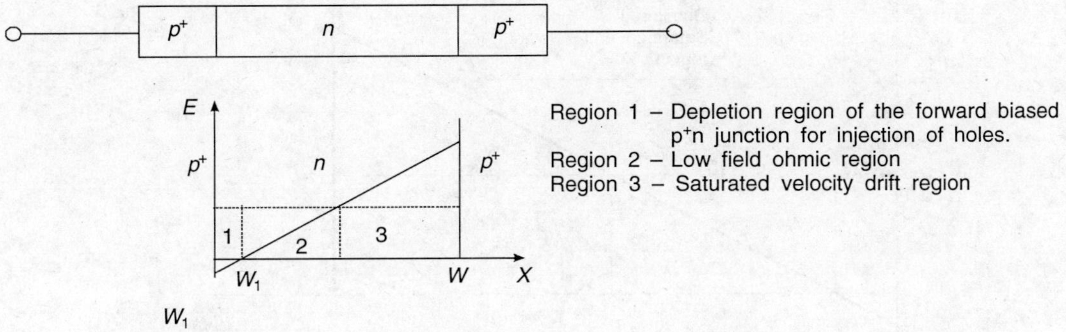

Figure 5.3 Field profile of a p^+np^+ BARITT device at reach through voltage.

When the depletion layer of the reverse biased junction reaches that of the forward biased junction, thermionic injection of minority carriers into the central n-region takes place across the barrier of the forward biased junction. The reverse field sweeps the injected carriers through the depletion layer of the reverse biased junction to the other terminal. When the applied voltage exceeds the punch through voltage, the forward biased junction will obtain a positive bias and current will begin to flow. The *V-I* characteristics are shown in Figure 5.4. The current below punch through is normal reverse saturated current.

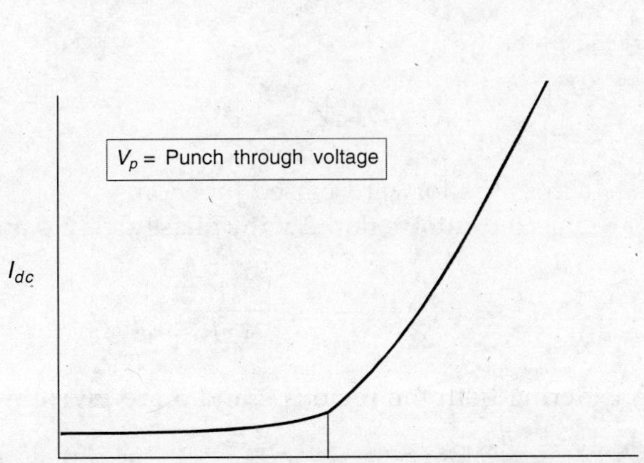

Figure 5.4 *I–V* characteristic of a BARITT diode.

5.2 SMALL SIGNAL MICROWAVE PROFILES OF A BARITT DIODE

Consider the simplified field profile shown in Figure 5.5 of different regions of a BARITT diode biased above punch through voltage. The region 1 is the forward biased p-n junction, which is equivalent to a parallel RC network,

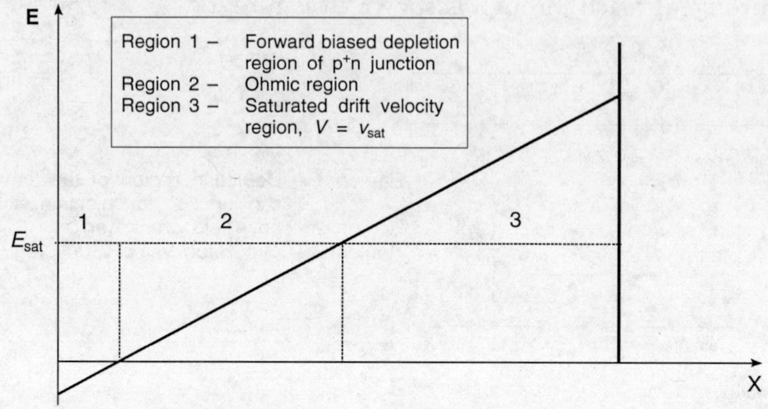

Figure 5.5 Electric field in different regions in p^+np^+ BARITT.

where

$$R = \frac{kT}{qI_0} \tag{5.4}$$

and C = The geometrical capacitance

$$C_1 \text{ (capacitance of the forward junction)} = \frac{\varepsilon A}{l_1} \tag{5.5}$$

l_1 = Width of the region 1

The total ac current is given by

$$I = j\omega C_1 V + \frac{V}{R} \tag{5.6}$$

where V is the ac voltage across the forward biased junction.

Thus the carriers are injected into region 2 with phase delay ϕ and is given by

$$\exp(-j\phi) = \frac{V}{RI} = \frac{1}{1 + jRC_1} \tag{5.7}$$

Thus the impedance considering both the regions 2 and 3 are given by

$$Z = \frac{1}{j\omega C_1}\left[1 - \frac{1 - \exp(-j\theta_1)}{j\theta_1(1 + j\omega RC_1)}\right] + \frac{1}{j\omega C_2}\left[1 - \frac{\exp(-j\theta_1)\{1 - \exp(-j\theta_2)\}}{j\theta_2(1 + j\omega RC_1)}\right] \tag{5.8}$$

where θ_1 is the transit angle of the ohmic region $\frac{wl_2}{v_s}$

$$C_2 = \frac{\varepsilon A}{l_2}, \qquad \theta_2 = \frac{wl_3}{v_s}$$

where l_2 and l_3 are widths of the regions 2 and 3.

Figure 5.6 shows the plot of the simplified theoretical impedance for a BARITT structure having $w = 3.5$ µcm, $w_2 = 4.5$ µcm, $v_{s1} = 5.6 \times 10^6$ cm/s and $v_s = 7.2 \times 10^6$ cm/s. Snapp and Weisglas [3] have measured the microwave chip impedance of a BARITT diode. For a p^+np^+ structure with n region of about 7µm, the peak negative resistance is obtained at 6.5 GHz

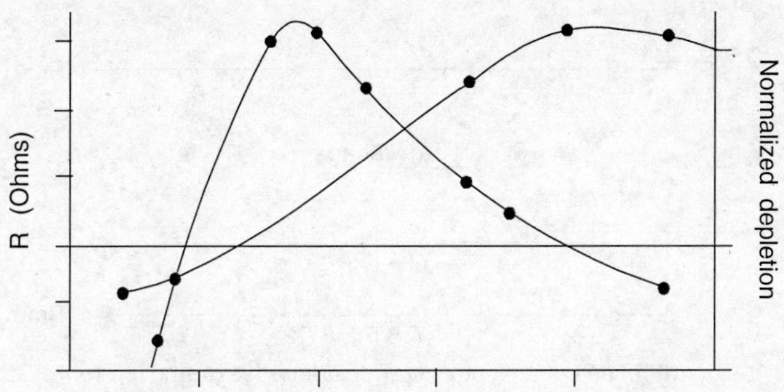

Figure 5.6 Small signal resistance and susceptance of a p^+np^+ BARITT device = 3×10^{-4} cm^2, $J = 25$ A/cm^2.

corresponding to a transit angle of 270°. The diode develops a negative microwave resistance at low current after punch through. As the current increases the peak negative resistance moves to higher frequencies. Further increase in current leads to decrease in the peak negative resistance. A detailed study of the small signal impedance of p^+nn^+ BARITTs have been carried out by Weisglas with current density as a parameter. The result indicates that the peak negative resistance reaches its maximum value at a rather low dc current of 50 A/cm^2 for n-region of 7 µm length and decrease at higher dc current densities. The resistance of the forward biased p^+n junction decreases with increasing dc current and tends to short circuit the junction capacitively at high current. Thus the high frequency negative resistance begins to decrease at lower current levels for BARITT structures, which leads to low output power and low efficiency.

5.3 LARGE SIGNAL OPERATION OF BARITT DIODE AND OPTIMUM FREQUENCY OF OSCILLATION

The large signal operation of a BARITT diode is shown in Figure 5.7. The carriers are thermionically injected over the barrier as delta function with the ac field and voltage reaches its maximum. Thus the peak current is delayed with respect to the ac voltage by a quarter of the time period $T/4$. The external current induced in this circuit when the charge bunch

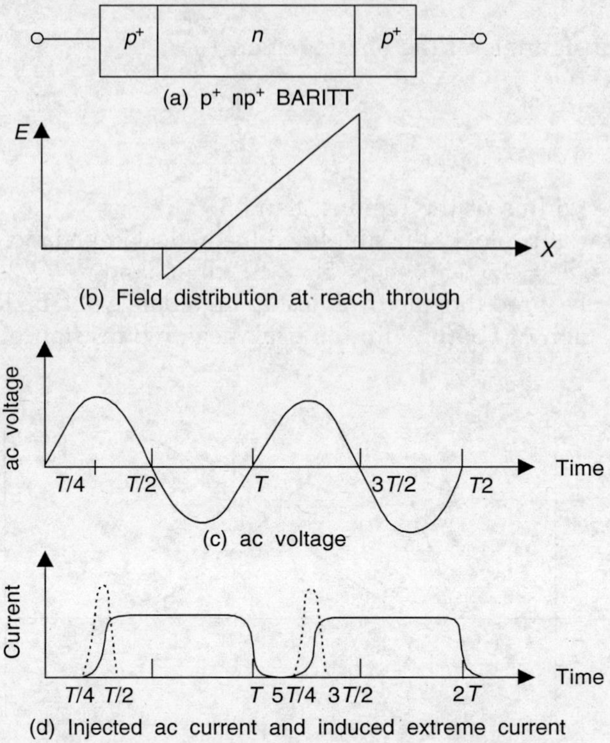

Figure 5.7 Operation of a BARITT diode subjected to a large signal ac voltage.

travels through the reverse biased depletion layer takes 3/4 of the time period to reach the negative terminal.

Thus we have

$$2\pi f \tau_d = 3\pi/2 \tag{5.9}$$

or

$$2\pi \frac{1}{T} \tau_d = 3\pi/2$$

Thus

$$\tau_d = \frac{3}{4} T = \frac{3}{4f} \tag{5.10}$$

or

$$f = \frac{3}{4\tau_d} \tag{5.11}$$

Now the drift velocity in a BARITT diode changes in the depletion region of the drift region as shown in the Figure 5.8.

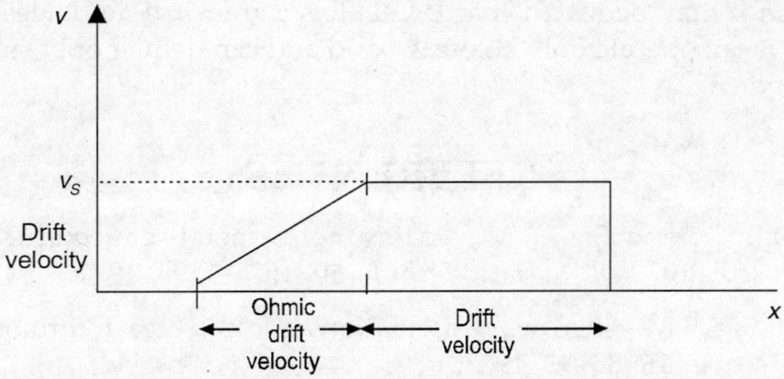

Figure 5.8 Drift velocity distribution in a BARITT device.

Transit time in the drift region is thus given by

$$\tau_d = \int_{x_R}^{x_s} \frac{dx}{\mu_n \mathcal{E}(n)} + \int_{x_s}^{w} \frac{dx}{v_s}$$

$$= \int_{x_R}^{x_s} \frac{dx}{\mu_n q N_D (x/\varepsilon_s)} + \frac{w - x_s}{v_s}$$

$$= \int_{x_R}^{x_s} \frac{\varepsilon_s}{q\mu_n N_D} \frac{dx}{x} + \frac{w - x_s}{v_s} \tag{5.12}$$

An approximate expression for f can be offered by assuming transit time to be w/v_s, and oscillation frequency is then given by

$$f = \frac{3v_s}{4w} \tag{5.13}$$

5.4 DISCUSSION

Microwave power has been observed for BARITT structures in the frequency range of 4–8 GHz. The power output varies from 2–8 mW for the mnm diodes reported by Coleman and Sze [1] operating between 4 and 6 GHz. An output power of 152 mW has been observed in Pt. Schottky BARITT diodes at 8.6 GHz with a peak efficiency of 2.3% [4]. The maximum efficiency obtained for a n^+ I p v n^+ BARITT diode has been found to be 5% with output power of 40 mW for a frequency of 7.6 GHz [5]. Unlike the IMPATT, the BARITT is a low noise device, since no random avalanche multiplication is involved in it. The noise arises only from the shot noise of injected carriers due to the random fluctuations of the velocity of the carriers in the drift zone. An experimentally measured noise figure up to 10–14 dB have been observed in a Si BARITT diode for a current density of 17 A/cm^2 at around 7 GHz. In BARITT diode the field is low over a considerable fraction of the drift region where the carriers move with non-saturated drift velocity determined by the low mobility. Thus the transit time is high and the field swing is limited, leading to low frequency and low power output of BARITTs.

In conclusion it may be stated that BARITT is a transit time device with good noise performance, but it can operate only below X-band and can deliver only small output power with low efficiency.

REFERENCES

1. Coleman, D.J., Jr. and Sze, S.M., A low noise metal-semiconductor-metal (MSM) microwave oscillator, *Bell System. Tech. J.*, **50**, 1675–1695, 1971.
2. Chu, J.L. and Sze, S.M., Microwave oscillations in $p^+n\ p^+$ reach through BARITT diode, *Solid State Electron*, **16**: 85–91, 1973.
3. Snapp, C.P. and Weisglas, P., *IEEE Transactions Electrons Devices*, **ED-19**, 1109, 1972.
4. Ahmed, S. and Freyer, J., High power Pt Schottky BARITT diode, *Electron. Lett.*, **12**: 238–239, 1975.
5. Konyan, O.E., Sze, S.M., and Yang, E.S., Microwave BARITT diode, *Solid State Electronics*, **19**, p. 795, 1976.

SUGGESTED READINGS

1. Liao, S.Y., *Microwave Device and Circuits*, Englewood Cliffs, NJ: Prentice Hall, 1990.
2. Weisglas, P., Avalanche and barrier injection devices *in* M.J. Howes and D.V. Morgan (Eds.), *Microwave Devices*, Wiley, 1976, Chap. 3.
3. Gibbons, G., *Avalanche-Diode Microwave Oscillators*, Clarendon Press, Oxford, 1973.

PROBLEMS

1. Compare the physical mechanism of carrier injection into the drift layer of a BARITT diode with that in IMPATT diode.
2. Compare the *I–V* characteristics of a BARITT diode with the same for an IMPATT diode and explain the difference between punch through voltage and the breakdown voltage.
3. Explain clearly the meaning and significance of the punch through voltage in a BARITT diode. Derive an expression for the same.
4. Find the frequency of oscillation in a silicon p^+np^+ BARITT diode. Given v_{sat} for silicon = 10^5 m/sec and width of the n-region = 7 microns.
5. Why the frequency of operation of a BARITT diode is limited to lower microwave frequencies and why the noise in BARITT is much smaller than in IMPATTs?

Chapter 6

Transfer Electron Devices (Gunn Diode)

6.0 INTRODUCTION

In 1961–62, Ridely, Watkins and Hilsum [1-2] showed a negative differential mobility (i.e., a decrease in velocity with the increase in field) of electrons from some compound semiconductor materials like GaAs, InP. This phenomenon is caused by the field-induced transfer of conduction band electrons from a low energy, high mobility valley to high energy, low mobility satellite valley. This phenomenon is known as "Transferred electron effect" or "Ridely-Watkins-Hilsum effect". The devices based on this effect are called "transferred electron devices".

The first experimental microwave oscillation based on this effect was observed by Gunn [3] in 1963 from an n-type GaAs and so the transferred electron devices were named as Gunn devices after his name. Later on devices based on InP, GaAsP, CdTe, and ZnSe have also exhibited this effect.

6.1 THE TRANSFERRED ELECTRON MECHANISM

The band structure diagram of GaAs is as shown in Figure 6.1, where electron energy is plotted against the propagation vector K. At room temperature and zero bias condition the

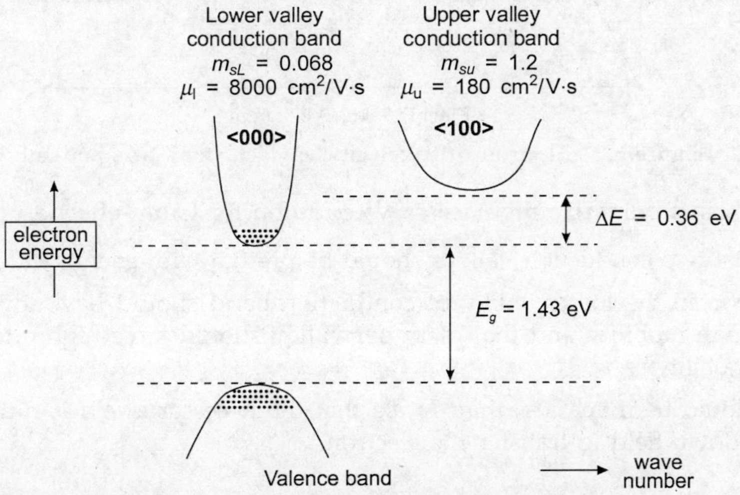

Figure 6.1 Simplified band structure of GaAs.

carriers are distributed between two valleys. But the thermal energy KT_0 of lower valley electrons at room temperature are much less than the energy difference ΔE between lower and upper valley of the conduction band and as a result no conduction/transfer of electrons from lower valley to upper valley takes place.

With the increase of biasing voltage the electrons get accelerated under the electric field and as a result their kinetic energy increases. The probability of this inter valley transfer of electrons is high as there are many available states in the upper valley. Thus more electrons from lower valley will move towards the upper valley, causing a redistribution of electrons between two valleys of the conduction band. Initially the lower valley electrons have a high value of mobility but due to transfer to the upper valley, mobility of lower valley electrons of the conduction band decreases and the effective mass increases. As a result current density J decreases which results in a negative differential conductivity/mobility. This negative conductivity effect depends only on the bulk properties of the semiconductor and not on junction or surface effects. It is therefore called a bulk negative differential conductivity (BNDC) effect.

Gunn observed that the carrier drift velocity, v_d increased linearly from zero to critical value of the electric field and beyond the threshold value of 3 kV/cm for the n-type of GaAs, the drift velocity decreased and the diode exhibited negative resistance. The dependence of electron drift velocity on electric field for GaAs and InP is shown in Figure 6.2.

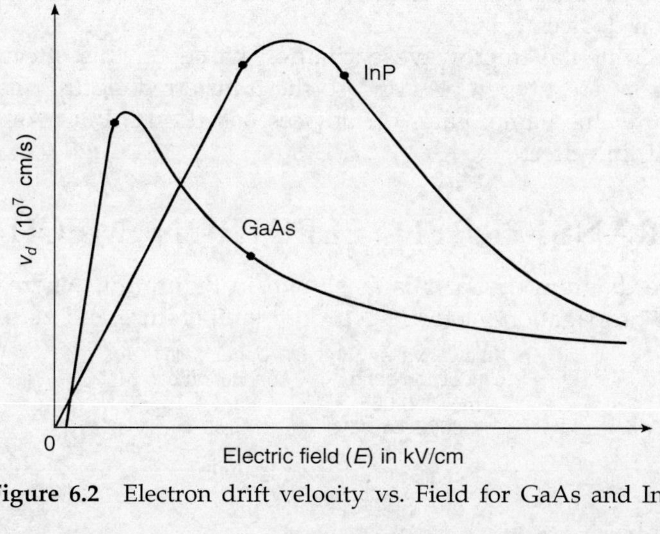

Figure 6.2 Electron drift velocity vs. Field for GaAs and InP.

Thus we can summarize the necessary condition for Gunn effect as:

1. At least two conduction valleys should be present with energy difference $\Delta E > KT_0$.
2. Electron in the lower valley of conduction band should have low effective mass and high mobility and that of upper valley should have high effective mass and low mobility.
3. ΔE should be much less than E_g, so that the necessary field is much less than the breakdown field to transfer the electron.

If m_l, μ_l, and n_l be the effective mass, mobility and carrier concentration respectively for lower valley, and m_u, μ_u, and n_u are the respective quantities for the upper valley, the steady state conductivity is given by

$$\sigma = q(\mu_l n_l + \mu_u n_u) \tag{6.1}$$

where q is the charge of the carrier.

Now the total carrier concentration $n = n_u + n_l$

When electrons are accelerated under high electric field E their effective temperature rises above the lattice temperature. Thus electron density n and mobility μ_u are both functions of electric field *(E)*.

Differentiating Equation (6.1) w.r.t. E

$$\frac{d\sigma}{dE} = q\left[\mu_l\left(\frac{dn_l}{dE}\right) + \mu_u\left(\frac{dn_u}{dE}\right)\right] + q\left[n_l\left(\frac{d\mu_l}{dE}\right) + n_u\left(\frac{d\mu_u}{dE}\right)\right] \tag{6.2}$$

Again, differentiating n w.r.t. E

$$\left(\frac{d}{dE}\right)(n_l + n_u) = \frac{dn}{dE} = 0$$

Because total number of carriers cannot change with electric field.

Hence

$$\frac{dn_l}{dE} = -\frac{dn_u}{dE} \tag{6.3}$$

Again the mobility

$$\frac{d\mu}{dE} \propto \left(\frac{d}{dE}\right)(E^p) = p.E^{p-1} = \frac{pE^p}{E} = \frac{\mu p}{E} \tag{6.4}$$

Substituting the value of Equations (6.3) and (6.4) in Equation (6.2)

$$\frac{d\sigma}{dE} = q(\mu_l - \mu_u)\left(\frac{dn_u}{dE}\right) + q(\mu_l n_l + \mu_u n_u)\frac{p}{E} \tag{6.5}$$

Again from Ohm's law the current density $J = \sigma E$; and differentiating J w.r.t. E gives

$$\frac{dJ}{dE} = \sigma + E\left(\frac{d\sigma}{dE}\right)$$

or

$$\left(\frac{1}{\sigma}\right)\left(\frac{dJ}{dE}\right) = 1 + \frac{\frac{d\sigma}{dE}}{\frac{\sigma}{E}} \tag{6.6}$$

As mentioned earlier for negative resistance to occur the current density (J) should decrease with the increase of electric field (E), i.e., dJ/dE is negative which is possible only when the right hand side of the Equation (6.6) is less than zero.

$$1+\frac{\left(\frac{d\sigma}{dE}\right)}{\left(\frac{\sigma}{E}\right)}<0$$

or

$$-\frac{\left(\frac{d\sigma}{dE}\right)}{\left(\frac{\sigma}{E}\right)}>1 \tag{6.7}$$

Dividing Equation (6.5) by Equation (6.1)

$$\frac{\left(\frac{d\sigma}{dE}\right)}{\sigma}=\frac{\left[q(\mu_l-\mu_u)\left(\frac{dn_u}{dE}\right)+q\left(\frac{p}{E}\right)(n_l\mu_l+n_u\mu_u)\right]}{[q(\mu_l n_l+\mu_u n_u)]}$$

writing $\frac{n_u}{n_l}$ = K and multiplying both sides by E we get

$$-\frac{\left(\frac{d\sigma}{dE}\right)}{\left(\frac{\sigma}{E}\right)}=-\left[\left\{\frac{(\mu_l-\mu_u)}{(\mu_l+\mu_u K)}\right\}\left\{\left(\frac{E}{n_l}\right)\left(\frac{dn_l}{dE}\right)\right\}+p\right]$$

From Equation (6.7)

$$\left[\left\{\frac{(\mu_l-\mu_u)}{(\mu_l+\mu_u K)}\right\}\left\{\left(-\frac{E}{n_l}\right)\left(\frac{dn_l}{dE}\right)\right\}-p\right]>1 \tag{6.8}$$

To maintain the inequality of Equation (6.8), $\mu_l > \mu_u$ and dn_l/dE should be negative. This proves that the electron at the lower valley has higher mobility and density of lower valley electron decreases with increase of electric field, which means that the electrons get transferred to upper valley. Now elemental semiconductors like silicon or germanium have only one valley in the conduction band and hence this effect is not possible. Again the electrons have lower effective mass and higher mobility than that of holes, thus n-type GaAs is suitable. It is to be noted here that inter-valley transfer must take place at realistic field levels. The threshold electric field at the same time should not raise the temperature so high as to cause significant generation of electrons from valence band by impact ionization, i.e., the applied threshold field should be much less than the breakdown field of the device.

6.2 DOMAIN FORMATION

A small applied voltage across the Gunn diode creates a uniform electric field, which results in uniform current density throughout the diode. At low voltage the GaAs is ohmic, since the drift velocity of the electron is proportional to the electric field. When the applied voltage is more than the threshold voltage the charge densities and the electric field within the sample become non-uniform creating domains, i.e., electrons in some region of the sample will be first to experience the field during inter-valley transfer. A high field domain (dipole) is formed near the cathode that reduces the electric field in rest of the material and causes the current to drop to about two-thirds of its maximum value.

Due to the formation of domain (accumulation of charge) an electric field is created as shown in Figure 6.3. It is to be noted that the carriers entering into the newly built field is more than those flowing out of the developed field. This high field domain then drifts with

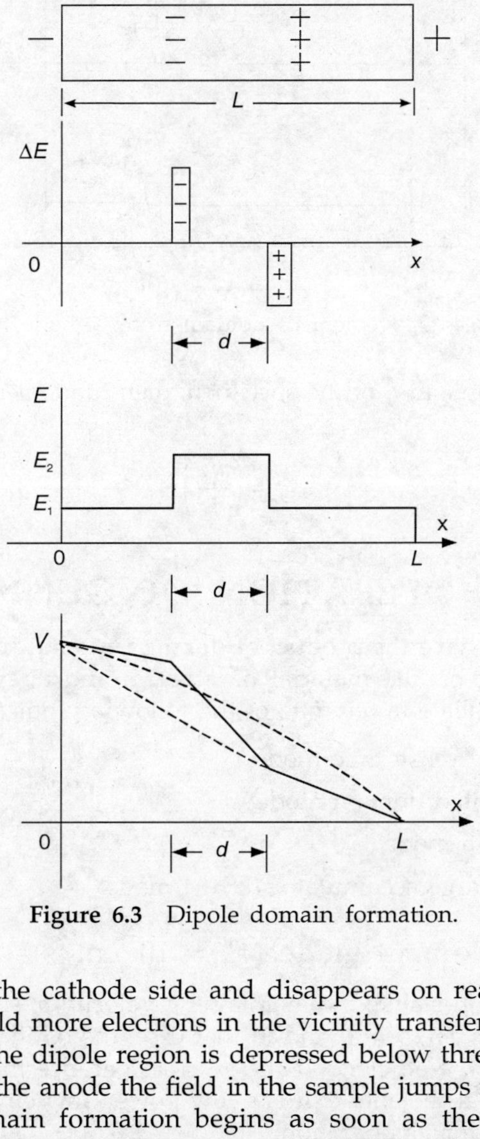

Figure 6.3 Dipole domain formation.

the carrier stream from the cathode side and disappears on reaching the anode. As the dipole drifts along the field more electrons in the vicinity transfer to the upper valley until the electric field outside the dipole region is depressed below threshold electric field. When the domain is reached to the anode the field in the sample jumps immediately to its original value and the next domain formation begins as soon as the field value exceeds the threshold value and the process is then repeated cyclically. The movements of a domain from cathode to anode are as shown in Figure 6.4. If v_d be the drift velocity of the domain

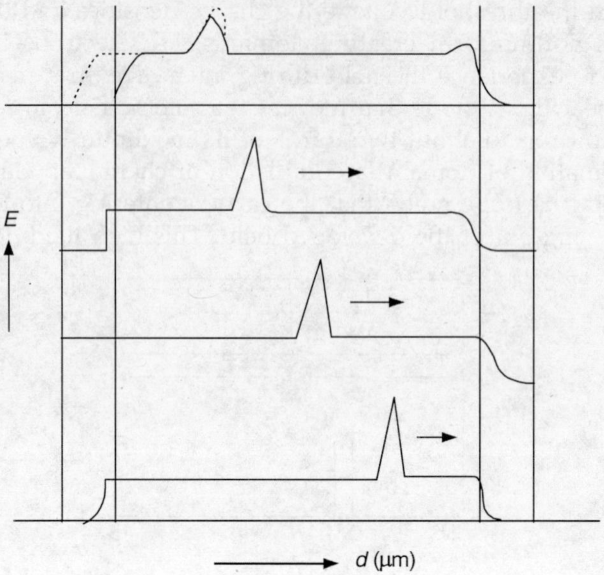

Figure 6.4 Movement of domain from cathode to anode.

with which it crosses the length L of the specimen, then fundamental frequency of operation is given by

$$f = \left(\frac{v_d}{L}\right) \tag{6.9}$$

6.3 MODES OF OPERATION OF GUNN DIODE

Most of the Gunn diodes have the product of doping (n_0) and drift lengths (L) greater than $10^{12}/\text{cm}^2$. But depending on the material parameters and operating conditions, a Gunn diode can be made to oscillate in any one of the following four frequency modes:

(a) Domain mode or transit time mode

(b) Delayed or inhibited domain mode

(c) Quenched domain mode

(d) Limited space-charge accumulator (LSA) mode.

6.3.1 Transit time domain mode ($fL = 10^7$ cm/s)

To make high field domain stable, the maintaining velocity v_s should be equal to the drift velocity v_d. Here $v_d = 10^{7}$ cm/s $= fL$ and in that case, the oscillation period = transit time; i.e., $\tau_o = \tau_t$. In this mode the efficiency is below 10% because the current is collected only when the domain arrives at the anode. It is a low power, low efficiency mode and does not need any external circuit for its operation.

6.3.2 Delayed domain mode (10^6 cm/s < fL < 10^7 cm/s)

Here the transit time is so chosen that the domain is collected when the applied field is less than critical field (E_{th}). The next domain can only form when the field reaches again the threshold value. In this case the oscillation period is greater than the transit time of the critical field (E_{th}). The next domain can only form when the field reaches again the threshold value. In this case the oscillation period is greater than the transit time, i.e., $\tau_o > \tau_t$. The efficiency of this mode is about 20%. This mode is also known as inhibited mode.

6.3.3 Quenched domain mode ($fL = 10^7$ cm/s)

During the negative half cycle, the bias field drops below the field (E_s). The domain collapses before it reaches the anode. When the bias field swings back again above threshold, a new domain is nucleated and the process repeats. Here, instead of transit time frequency, the device oscillates at resonant frequency of the circuit, which is several times the transit time frequency. The threshold efficiency of the oscillator in this mode is up to maximum of 13%.

6.3.4 Limited space-charge accumulation (LSA) mode ($fL > 2 \times 10^7$ cm/s)

At a very high frequency due to the shortage of time, domain cannot form but the field is above threshold. Due to that most of the domain are maintained in the negative conductance state during a large fraction of the voltage cycle. Any accumulation of electrons near the cathode has time to collapse while the signal is below threshold. At this field level the current in the device is then proportional to the drift velocity. The device operates in LSA mode with an efficiency of 20%. Here the oscillation period (τ_o) is few times of the dielectric relaxation time (τ_d). The lower and upper limit of the drift velocity are 5×10^6 to 5×10^7 cm/s and the product of doping and length (n_0L) is nearly $10^{12}/\text{cm}^2$. Thus the ratio of n_0/f within the above mentioned limit is 2×10^5 to 2×10^4. All the above-mentioned modes of Gunn diode are shown in Figure 6.5.

Problem 6.1 A Gunn diode is working in transit time modes at 12 GHz. The domain of charges move at a speed of 10^7 cm/s. Calculate (i) the length of the device, (iia) can the device work at 10 GHz and (iib) 14 GHz? (iii) Which is the mode of operation in each case.

Solution If L be the length of the device, τ_o be the transit time of the carrier, then saturated drift velocity

$$v_s = \frac{L}{\tau_o}$$

or

$$\text{(i) } L = v_s \times \tau_o = \frac{10^7}{12 \times 10^9} = 8.33 \times 10^{-4} = 83.3 \text{ mm}$$

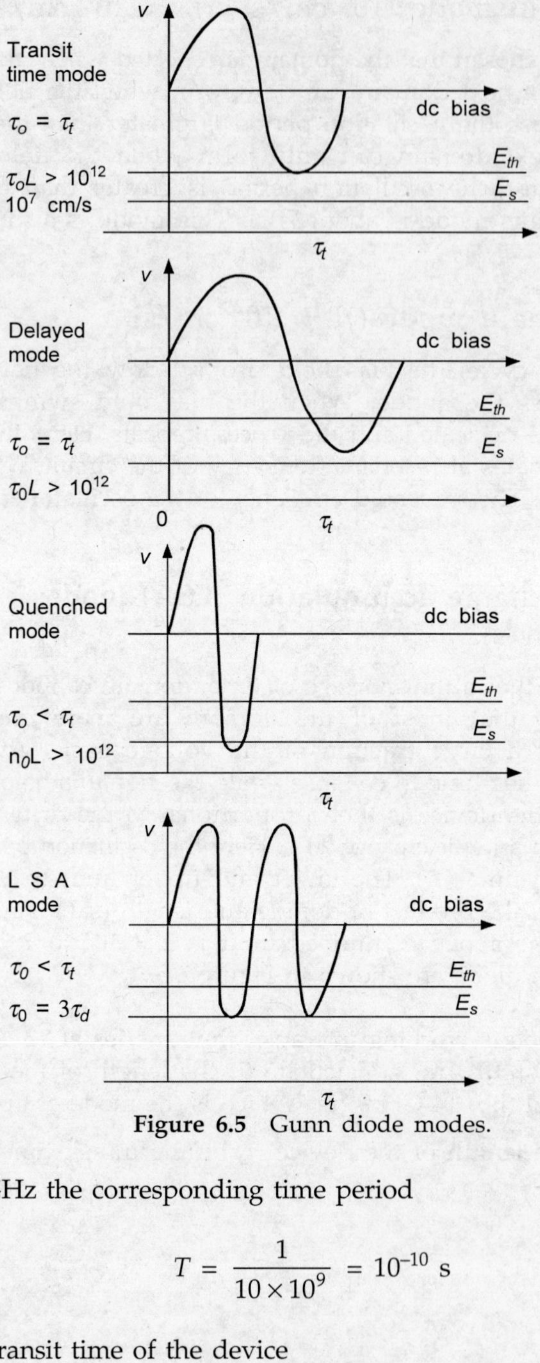

Figure 6.5 Gunn diode modes.

(iia) For 10 GHz the corresponding time period

$$T = \frac{1}{10 \times 10^9} = 10^{-10} \text{ s}$$

Whereas the transit time of the device

$$\tau_o = 0.08 \times 10^{-9} \text{ s}$$

As $T > \tau_o$, it can operate in delayed mode.

(iib) For 14 GHz, the corresponding time period is

$$T = \frac{1}{14 \times 10^9} = 0.07 \times 10^{-9} \text{ s}$$

(iii) As $T < \tau_o$, hence it can operate in quenched mode.

Problem 6.2 Design a Gunn diode which operates at 10 GHz, having $n_0 = 4 \times 10^{15}$ cm^{-3}, $E_{th} = 3$ kV/cm, $E_s = 20$ kV/cm; and it can generate 100 mW *rf* output power with 3% conversion efficiency.

Solution: Here $f = 10$ GHz and $v_s = 10^7$ cm/s

The device length is $L = \dfrac{v_s}{f} = 10^{-3}$ cm

The threshold voltage is

$$V_{th} = E_{th}L = 3000 \times 10^{-3} = 3 \text{ V}$$

Generally the operating voltage is approximately three times the threshold voltage.

Hence $$V_{bias} = 3\,V_{th} = 3 \times 3 = 9 \text{ V}$$

Since $\eta = 0.03$, $P_0 = 100$ mW

Hence $$P_{dc} = \frac{P_0}{0.03} = \frac{(100 \times 10^{-3})}{0.03} = 3.3 \text{ W} \quad \left(\because \frac{P_o}{P_{dc}} = \eta\right)$$

Thus the amount of dc input power will be dissipated as heat

$$= P_{dc} - P_0 = 3.2 \text{ W}$$

$$I_{bias} = \frac{P_{dc}}{V_{bias}} = 3.3/9 = 0.367 \text{ A.}$$

Hence current density $J = n_0 q v_s = (4 \times 10^{15})(1.6 \times 10^{-19}) \times (10^7)$

$$= 6.4 \times 10^3 \text{ A/cm}^2 = \frac{I_{bias}}{A}$$

where

$$A = \text{Cross-sectional area of the diode}$$

$$= \pi r^2 = \frac{I_{bias}}{J} = 0.367/(6.4 \times 10^3)$$

Hence $$r = \frac{(0.367 \times 10^{-3})}{(6.4 \times 3.14)} = 0.0042661 \text{ cm}$$

6.4 CRITERION FOR CLASSIFYING THE MODES OF OPERATION

When a Gunn diode is biased at early stages of space charge accumulation, their growth rate is given by:

$$Q(x,t) = Q(X - vt, 0)\exp\left(\frac{t}{\tau_d}\right) \tag{6.10}$$

Where the negative dielectric relaxation time

$$\tau_d = \frac{\varepsilon}{\sigma} = \frac{\varepsilon}{\left(\dfrac{en_0}{\mu_m}\right)}$$

where

n_0 = Doping concentration
σ = Conductivity
e = Charge of an electron
μ_m = Negative mobility of the n-type semiconductor
ε = Dielectric permittivity of the semiconductor.

The factor of maximum growth is given by:

$$\text{Growth factor} = Q\,\frac{\left(L, \dfrac{L}{v}\right)}{Q(0,0)}$$

$$= \exp\left(\frac{L}{v\tau_d}\right)$$

$$= \exp\left\{\frac{Ln_0e\,|\,\mu_0\,|}{(\varepsilon v)}\right\} \tag{6.11}$$

The space charge layer starts to develop at $t = 0$, $X = 0$ (cathode side) and reaches the anode at a distance $X = L$ after time $t = L/v$. This growth factor must be larger than unity for a large space charge growth and is possible when

$$n_0L > \frac{(\varepsilon v)}{(e\,|\,\mu_0\,|)} \tag{6.12}$$

More accurate numerical analysis gives the value of $n_0L > 10^{12}$ cm^{-3}. Devices which satisfy this inequality are called supercritical and those whose $n_0L < 10^{12}$ cm^{-3} are called subcritical.

Problem 6.3 The parameter of an n-type GaAs Gunn diode are given below:
Electron drift velocity = 2×10^5 m/s
Relative dielectric constant = 13.1

and negative electron mobility = $0.0145\ \dfrac{\text{m}^2}{\text{V}\cdot\text{s}}$

Determine the criterion for classifying the modes of operation.

Solution: From Equation (6.12) the criterion is

$$\frac{(\varepsilon v)}{(e\,|\,\mu_0\,|)} = \frac{(8.854\times10^{-12}\times13.1\times2\,\times\,10^{5}}{(1.6\times10^{-19}\times0.0145)}$$

$$= 9998.9\,\times\,10^{12}/\text{m}^2$$

$$=\;0.99\,\times\,10^{12}\;\text{cm}^2$$

6.5 MODE CHART FOR TRANSFERRED ELECTRON DEVICES

The operation at different modes mentioned under Section 6.3 depends on the value fL and n_0L and explained with the help of mode chart [4] as shown in Figure 6.6. From mode chart one can easily get an idea about the region of the chart which can be used for stable amplification purpose.

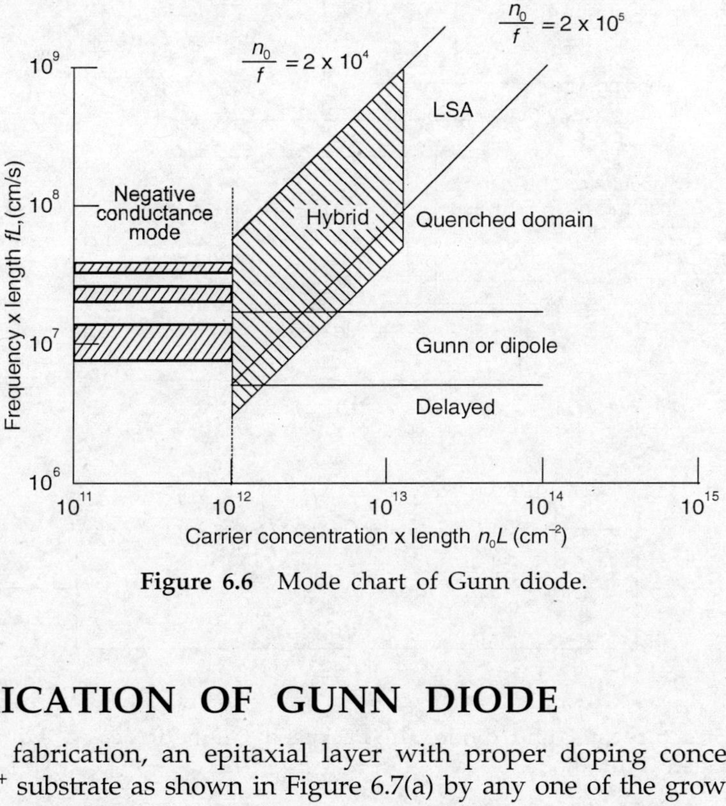

Figure 6.6 Mode chart of Gunn diode.

6.6 FABRICATION OF GUNN DIODE

In Gunn diode fabrication, an epitaxial layer with proper doping concentration (n_0) is grown on the n$^+$ substrate as shown in Figure 6.7(a) by any one of the growth methods like vapour phase epitaxy (VPE), liquid phase epitaxy (LPE) or molecular beam epitaxy (MBE). An n$^+$ buffer epitaxial layer is grown on the top of the n-layer for contact purposes. The structure of a Gunn diode is as shown in the Figure 6.7(a).

Then the fabricated device is mounted on top of a heat sink. The package is special pill type (S_4) and for each of the integration, the entire package is mounted on top of a screw.

Because of low conversion efficiency proper heat sink is very much essential. A packaged Gunn diode is as shown in Figure 6.7(b) and equivalent circuit of packaged Gunn diode is shown in Figure 6.7(c).

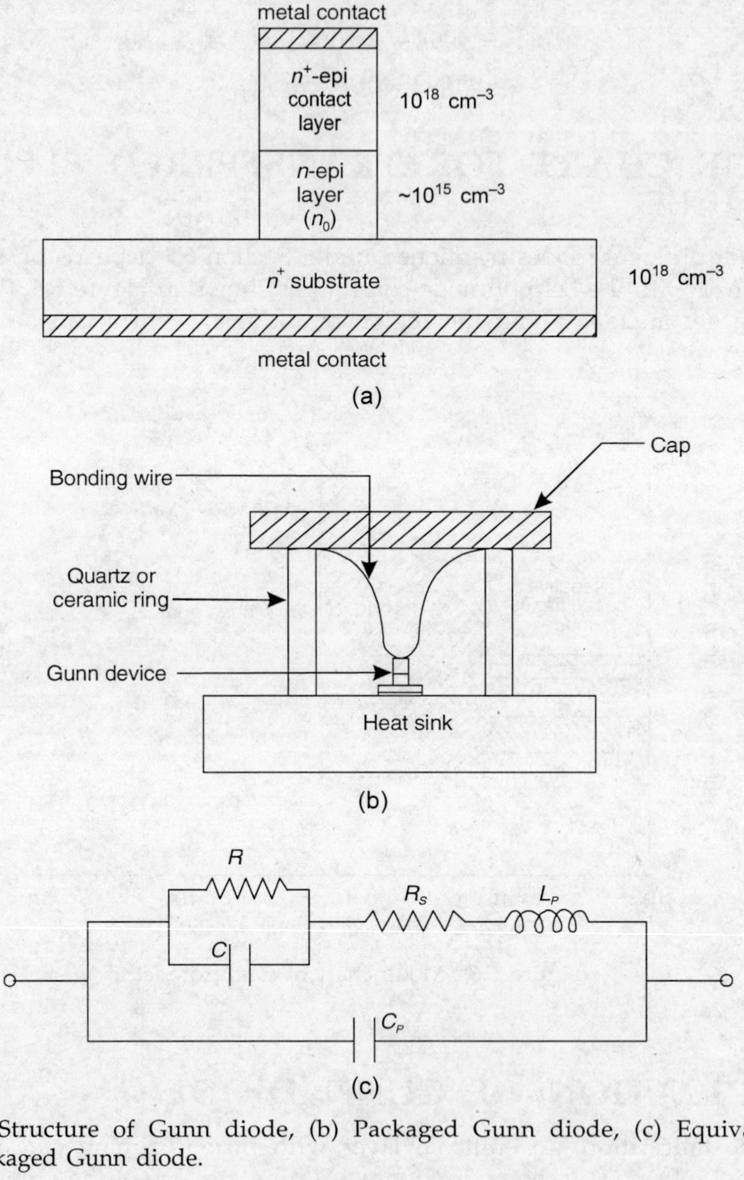

Figure 6.7 (a) Structure of Gunn diode, (b) Packaged Gunn diode, (c) Equivalent circuit of packaged Gunn diode.

6.7 TYPICAL CHARACTERISTICS

A typical biased voltage of 10–12 V with a bias current of 300 mA gives a power output of 100 mW in the X-band.

Typical frequency range : 4 to 94 GHz
Efficiency : 10–20%
CW Power output : 250 mW at 15 GHz
Pulsed power : 5 W at 12 GHz

6.8 APPLICATION OF GUNN DIODE

1. High power Gunn diodes are used in CW Doppler radar.
2. Frequently used as low and medium power oscillator in microwave receivers.
3. Broadband linear amplifiers.
4. Fast combinational and sequential logic circuits.
5. Because of low noise figure it can be used as a pump source for most of the parametric amplifiers.
6. Pulsed Gunn diode can be used in industry, telemetry systems and in transponder for air traffic control.

6.9 GUNN OSCILLATOR CIRCUITS

Gunn oscillator circuit can be designed in coaxial cable, wave-guide, micro strips in the frequency range of 8–94 GHz. In Figure 6.8 some of the common techniques to couple a packaged diode into a wave-guide circuit under biased post filter is shown. Here the post, coaxial section, and the reduced height wave-guide transformer are used to match the low

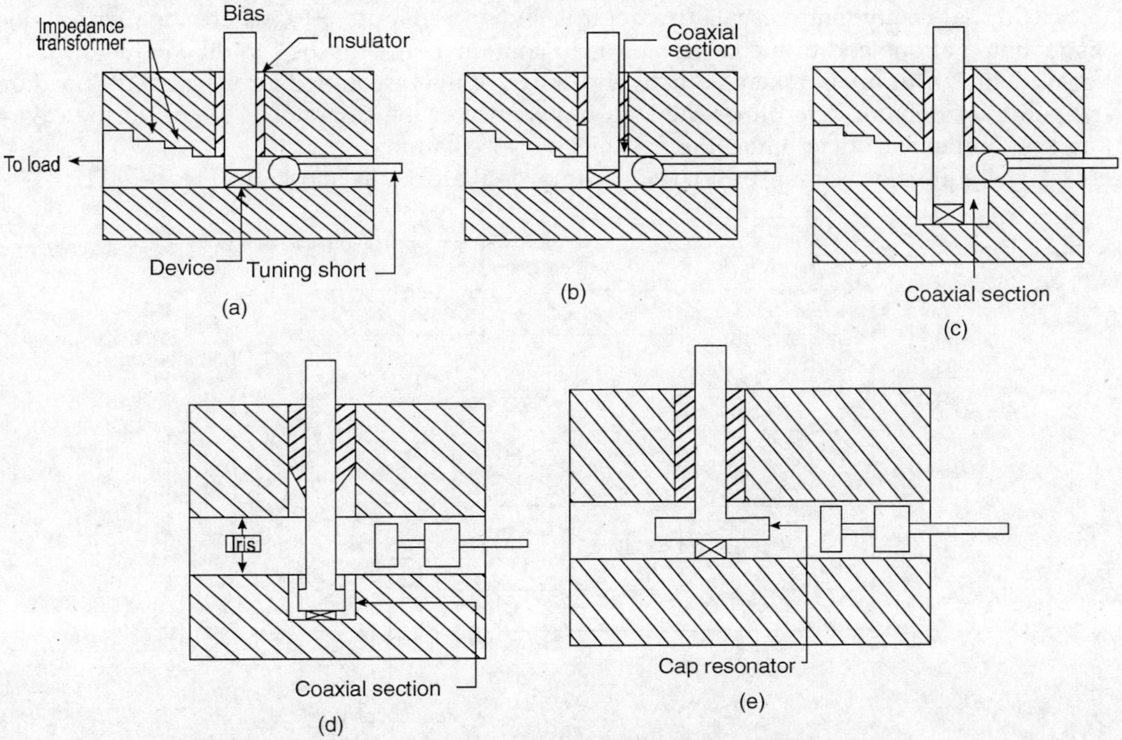

Figure 6.8 Wave-guide circuits for transferred electron devices.

device impedance to the high wave-guide impedance. Also the frequency of the oscillator can be controlled with the help of cap resonator (combination of disk and post). A sliding short tuner is used in all the circuits to obtain maximum output power from the oscillation at a desired frequency.

Theoretical and experimental analysis of a wave-guide mounted Gunn diode was given by Eisenhart [5] in 1971. A simplified circuit for a Gunn oscillator is as shown in Figure 6.9.

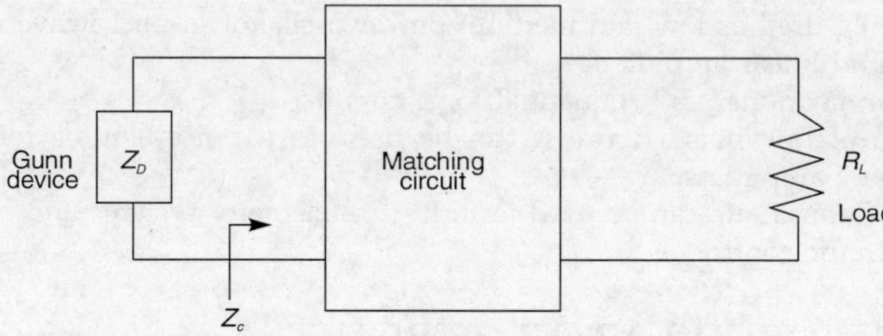

Figure 6.9 Simplified circuit for a Gunn oscillator.

The condition for oscillation

$$|Re(Z_D)| > Re(Z_c) \text{ and } |I_m(Z_D)| > Re(Z_c)$$

where Z_D and Z_c are device and circuit impedance respectively.

Coaxial cavity Gunn oscillator circuit is shown in Figure 6.10, where the Gunn diode is mounted at one end of the cavity and is in continuation with the central conductor of the coaxial line. Output is taken inductively or capacitively coupled probe and its position within the resonator determines the load impedance of the circuit. The length of the cavity and the impedance determines the frequency of oscillation.

The equivalent circuit of a coaxial Gunn oscillator is as shown in Figure 6.11.

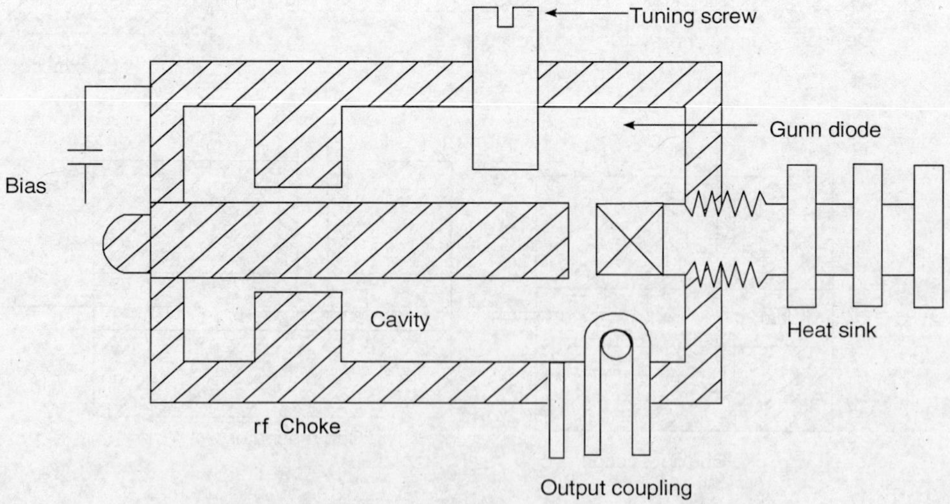

Figure 6.10 Gunn oscillator using coaxial cavity.

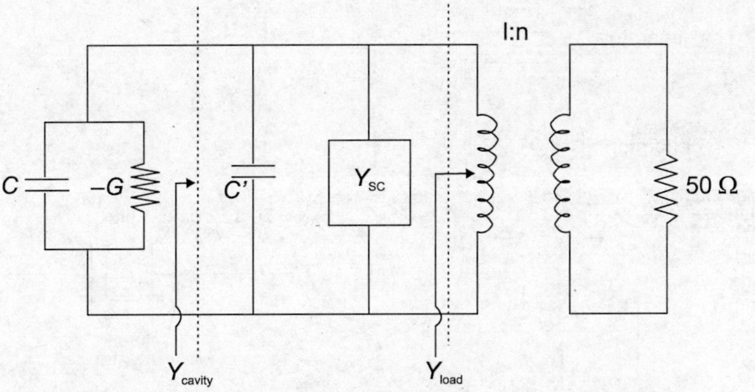

Figure 6.11 Equivalent circuit of coaxial Gunn oscillator.

Here the capacitor of the tuning screws is used to tune the device. *B* and *G* are the susceptance and conductance of the device and cavity circuit. The oscillation condition is to be satisfied

$$B_{\text{device}} + B_{\text{circuit}} = 0 \tag{6.13a}$$

$$G_{\text{device}} + G_{\text{circuit}} = 0 \tag{6.13b}$$

From the equivalent circuit we can write

$$jwc - G + jwc' + Y_{s.c} + Y_L = 0 \tag{6.14}$$

where,

$$Y_{s.c} = -j \cot\left(\frac{2\pi L}{\lambda_0}\right) \tag{6.15}$$

and *L* is the length of the cavity.

Therefore the two conditions of oscillations are

$$wc + wc' - \cot\left(\frac{2\pi L}{\lambda_0}\right) + Im\,(Y_L) = 0 \tag{6.16a}$$

and

$$-G + Re(Y_L) = 0 \tag{6.16b}$$

Out of all these oscillator circuits the wave-guide type is most common and popular.

6.10 GUNN DIODE AMPLIFIER

Like other negative resistance devices Gunn diode can be used as an amplifier but are not very popular. Average gain ranges from 1–12 dB with a noise figure of 15 dB. The amplifier operates satisfactorily within a frequency range of 4–16 GHz. The basic circuit of Gunn diode amplifier is shown in Figure 6.12.

The gain of this amplifier can be obtained by replacing *G* by –*G* in the definition of the reflection coefficient of the one port circuit terminating port 3 of the circulator.

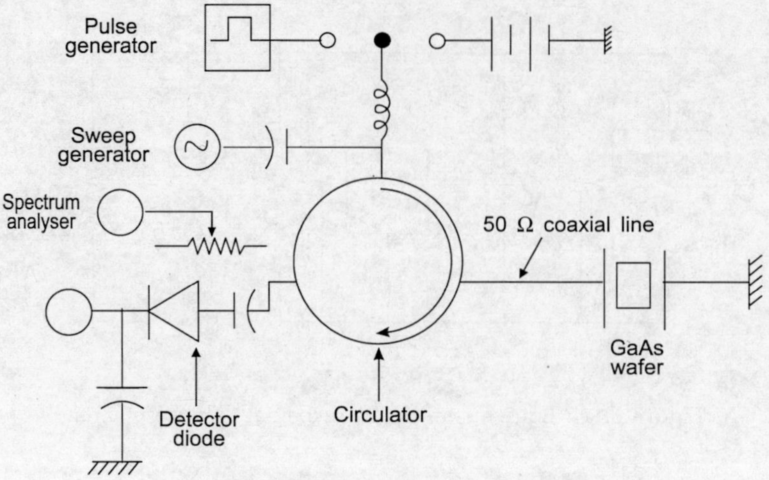

Figure 6.12 Gunn diode amplifier.

6.11 DISADVANTAGE OF GUNN DIODE

Gunn diode is very much temperature dependent i.e., a frequency shift of 0.5 to 3 MHz per degree centigrade. By proper design this frequency drift can be reduced to 50 kHz per degree centigrade for a range of – 40°C to 70°C.

6.12 InP GUNN DIODE

The principle of operation of n-type Indium Phosphide (InP) Gunn diode under dc biasing condition is same as that of GaAs Gunn diode. In the year 1970 Hilsum proposed that Indium Phosphide or some of the compounds of Indium cadmium antimonide can show the Gunn effect. In Figure 6.13 the three valley model of InP is shown.

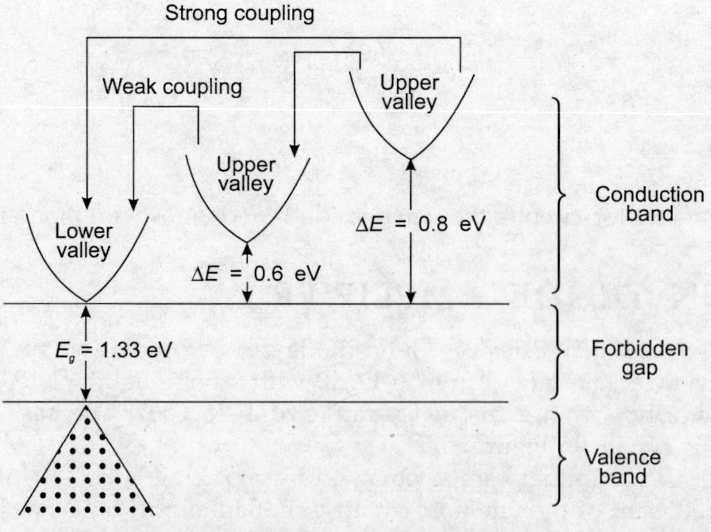

Figure 6.13 Three-valley model energy level for InP diode.

In InP Gunn diode rapid transfer of electrons from lower to upper valley with the increase of field makes larger peak to valley current ratio than GaAs Gunn diode as shown in Figure 6.13. For this process the middle valley energy level of InP provides the additional energy loss mechanism required to avoid breakdown caused by the high energies acquired by the lower-valley electrons which are weakly coupled with middle-valley but strongly to the upper-valley. Thus the situation ensures that concentration of electrons in the middle-valley is much more at normal operating conditions. Unlike the domain formation in GaAs Gunn diode, the three-valley model of InP inhibits the formation of domain because the electron diffusion coefficient is increased by the stronger coupling [6]. Also Taylor and Colliver [7] showed experimentally that InP Gunn diode oscillates in transit time mode but do not oscillate in the bulk mode like LSA. A wide frequency range of oscillation is possible by proper selection of epilayer thickness and by adjusting the cavity size. It has been reported that at a frequency of 5.5 GHz, 3.05 Watt pulse power is available with a conversion efficiency of 14.7% 8 and 0.2 Watt CW power at 18 GHz with a conversion efficiency of 10.2%.

REFERENCES

1. Ridely, B.K. and Watkins T.B., The possibility of negative resistance effects in semiconductors, *Proc. Physics Soc.*, **78**, 293–304, August 1961.

2. Hilsum, C., Transferred electron amplifiers and oscillators, *Proc., IEEE*, **50**, 185–189, February 1962.

3. Gunn, J.B., Microwave oscillations of current-in III-V semiconductors, *Solid State Communication*, **I**, 88–91, September 1963.

4. Uenohara, M., Bulk gallium arsenic devices Chapter 16 in H.A. Watson (Ed.) *Microwave Semiconductor Devices and their Circuit Application*, McGraw-Hill Book Company, New York, 1969.

5. Eisenhart R.L. and Khan, P.J., Theoretical and experimental analysis of a wave guide Mounting structure, *IEEE Transaction on Microwave Theory and Techniques*, Vol. MTT-19, August, 1971, pp. 706–709.

6. G.B.L., Three level oscillator in Indium phosphide, *Physics Today*, Vol. 23, pp. 19–20, December 1970.

7. Taylor, B.C and Colliver, D.J., Indium phosphide microwave oscillators, *IEEE Transaction on Electron Device*, ED-18, No-10, pp. 835–840, October 1971.

8. Colliver, D. and Prew, B., Indium phosphide is it practical for solid state microwave sources? *Solid State Electronics*, 110–113, April 10, 1992.

SUGGESTED READINGS

1. Sze, S.M., *Physics of Semiconductor Devices*, 2nd ed., John Wiley & Sons, New York, 1985.
2. Liao, S.Y, *Microwave devices and circuits*; 3rd ed., Prentice-Hall of India, New Delhi, 1994.
3. Eastman, L.F., *Gallium Arsenide Microwave Bulk and Transit Time Devices*, Artech House, Dedham, Mass., 1973.

PROBLEMS

1. With the help of two-valley theory explain the operation of a Gunn diode.
2. Describe the several domain formation modes of a Gunn diode.
3. Design a Gunn device operating at 10 GHz. The electrons travel at a saturated speed of 10^7cm/s.
4. Describe the fabrication and packaging of Gunn diode.
5. Design a GaAs Gunn diode to operate at 5 GHz in the Gunn mode, with doping concentration of $n_0 = 2 \times 10^5 \text{cm}^{-3}$, $E_s = 22$ kV/cm, $E_{th} = 3.5$ kV/cm and the device produces 150 mW of output power with 10% dc to rf efficiency.
6. Mention the advantages and disadvantages of Gunn diode.
7. Mention the typical characteristics and applications of a Gunn diode.
8. A Gunn diode is working in transit time mode at 10 GHz. Can the device work at 8–12 GHz? Which is the mode of operation in each case?
9. Explain how a Gunn diode can be used as an oscillator.
10. Why elemental semiconductors like silicon, germanium are not suitable for fabrication of Gunn diode?
11. Determine the criterion for classifying the modes of operation of an n-type GaAs Gunn diode having the following parameters
    Electron drift velocity = 3×10^5 m/s,
    Negative Electron mobility = 0.015 $\text{m}^2/\text{V}\cdot\text{s}$,
    Relative dielectric constant = 12.5.
12. Write down the difference between transferred electron devices and avalanche transit time devices.
13. For a transit time mode, the domain velocity is equal to the carrier drift velocity of about 10^5 m/s. Determine the drift length of the diode at a frequency of 10 GHz.
14. A typical Gunn diode has the following specification:
    Operating frequency: 10 GHz
    Device length: 100 mm
    Voltage pulsed amplified: 30 V
    Determine the threshold electric field.

15. Why Gunn diode amplifier is not so popular?

16. A transferred electron diode operates in transit time mode at 10 GHz. If the velocity of propagation of domain is taken as 10^5 m/s, calculate the length of the device.

17. A typical n-type GaAs Gunn diode has the following parameters:

Threshold field (E_{th}): 2.8 kV/cm
Applied field (E): 3.2 kV/cm
Active length of the device (L): 25 μm
Doping concentration (n_0): 10^{14} cm^{-3}
Operating Frequency (f): 12 GHz

(a) Compute the electron drift velocity.
(b) Calculate the current density.
(c) Estimate the negative electron mobility.

18. Describe the three-valley model of Gunn diode.

Chapter 7

Tunnel Diode

7.0 INTRODUCTION

In an ordinary p-n junction diode with the impurity doping of about 1 part in 10^8 creates a depletion layer of width 5-micron around the junction. This depletion layer constitutes a potential barrier at the junction which restrains the flow of carriers from the side of the junction where the majority carrier has higher concentration gradient to the side of lower concentration gradient. If the concentration of impurity atoms is greatly increased to 1 part in 10^3 (corresponding to a density in excess of 10^{19} cm^{-3} or more), the device characteristics are completely changed. Esaki announced this new diode in 1958 in his classic paper on tunnel diode [1]. Esaki described this anomalous phenomenon by applying quantum-tunnelling theory. The tunnelling phenomenon is a majority carrier effect.

7.1 TUNNELLING PROCESS

Figure 7.1(a) shows the energy band diagram of two isolated semiconductors with a distance d between them. The potential barrier height eV_0 is equal to the electron affinity ex. If the distance d is sufficiently small then the electrons in the left side semiconductor may transport across the barrier and move to the right side semiconductor, even if the electron energy is much less than the barrier height. This phenomenon is the quantum-tunnelling phenomenon.

Figure 7.1(a) is redrawn to get one-dimensional potential barrier of Figure 7.1(b). The particle will always be reflected if its energy E is less than the potential barrier energy eV_0. According to quantum theory, the particle has finite probability to transmit or tunnel through the potential barrier. With the help of Schrodinger's equation the behaviour of a particle in the region $eV(x) = 0$, thus

$$-\frac{\hbar^2}{2m^*}\frac{d^2\Psi}{dx^2} = E\Psi \tag{7.1}$$

or

$$\frac{d^2\Psi}{dx^2} = -\frac{2m^*E}{\hbar^2}\Psi \tag{7.2}$$

where m^* is the effective mass, $\hbar$ is the reduced Planck constant, Ψ is the wave function and E is the kinetic energy of the particle.

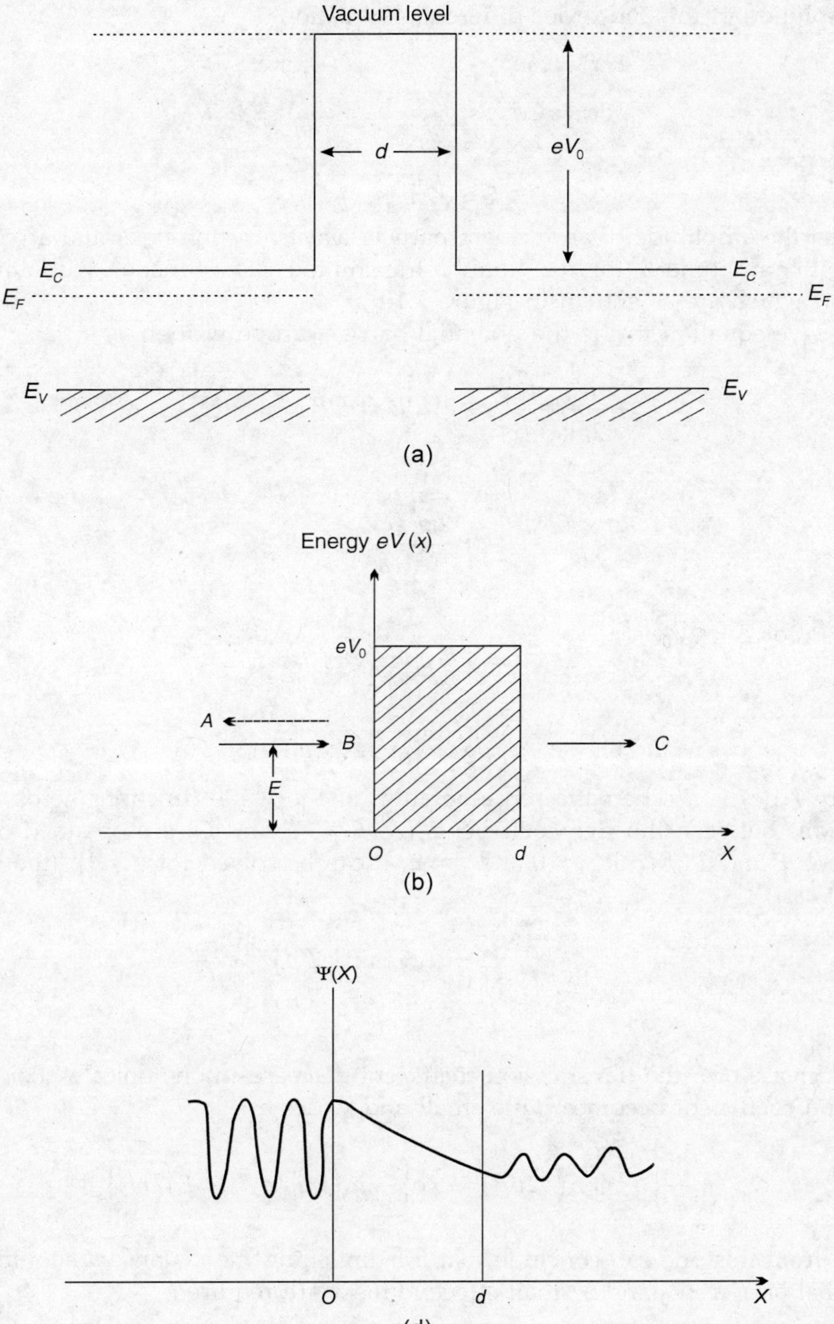

Figure 7.1 (a) Band diagram of two isolated semiconductors with a distance *d*, (b) one dimensional potential barrier, and (c) schematic representation of the wave function across the potential barrier.

The solution of the 2nd order differential equation

$$\Psi(x) = Ae^{jkx} + Be^{-jkx} \qquad \text{for } x \le 0 \tag{7.3}$$

$$\Psi(x) = Ce^{jkx} \qquad \text{for } x \ge d \tag{7.4}$$

where

$$k = \sqrt{2m^* E/\hbar^2}$$

for $x \le 0$, let the amplitude of an incident particle wave function be A and a reflected wave function with amplitude B; for $x \ge d$ only a transmitted wave function with amplitude C is present and which are as shown in Figure 7.1(b).

The wave equation inside the potential barrier can be written as

$$-\frac{\hbar^2}{2m^*}\frac{d^2\Psi}{dx^2} + eV_0\Psi = E\Psi \tag{7.5}$$

or

$$\frac{d^2\Psi}{dx^2} = \frac{2m^*(eV_0 - E)}{\hbar^2}\Psi \tag{7.6}$$

The solution for $E < eV_0$ is

$$\Psi(x) = Fe^{\beta x} + Ge^{-\beta x} \tag{7.7}$$

where

$$\beta \cong \sqrt{2m^*(eV_0 - E)/h^2}$$

Figure 7.1(c) is a schematic representation of the wave functions across the barrier. Four relations between the five coefficients (A, B, C, F and G) are obtained by using the continuity of Ψ and $d\Psi/dx$ at $x = 0$ and $x = d$. It can be solved for $(C/A)^2$ (the transmission coefficient) as

$$\left(\frac{C}{A}\right)^2 = \left[1 + \frac{(eV_0 \sinh \beta d)^2}{4E(eV_0 - E)}\right]^{-1} \tag{7.8}$$

with the decrease of E the transmission coefficient decreases monotonically. For $\beta d \gg 1$, the transmission coefficient becomes quite small and varies as

$$\left(\frac{C}{A}\right)^2 - \exp(-2\beta d) = \exp\left[-2d\sqrt{2m^*(eV_0 - E)/\hbar^2}\right]$$

Now from this one can conclude that for tunnelling the distance d should be small, a low potential barrier eV_0 and a small effective mass are required.

7.2 PRINCIPLE OF TUNNEL DIODE

Thus from the theory of quantum tunnelling phenomenon a p-n junction diode can act as a tunnel diode when it satisfies the following condition for tunnelling:

(i) Fermi level is within the band.

(ii) The barrier of the depletion layer at the junction is very thin (≅ 100A°) and which can be achieved by means of heavy doping (≅ 10^{19} to 10^{21} atoms/cc) in both the p and n-layer. Thus the particle has the lower KE and it cannot cross the barrier height but it has the probability that it can tunnel through the potential barrier.

(iii) At 0 K, all the states above Fermi level are empty and below Fermi level all the states are occupied.

(iv) There must be filled energy states on the side from which the particle will tunnel and allow empty states on the other side into which the particle penetrates through the same energy level.

Figure 7.2 shows the energy band diagram of a tunnel diode under no bias, forward bias or reverse bias condition. Initially at zero bias condition as no filled states

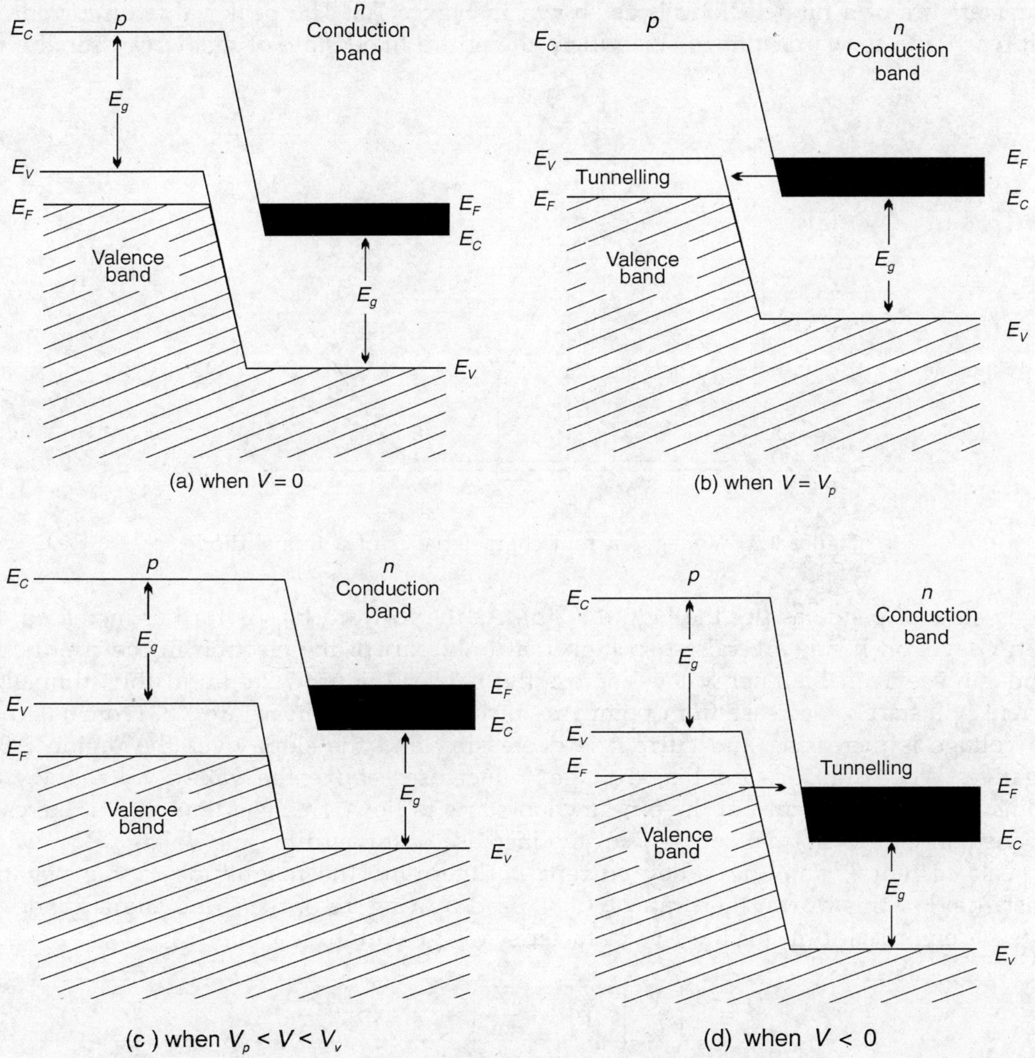

Figure 7.2 Energy band diagram of a Tunnel diode.

are available in any side of the junction, no tunnelling takes place. With the application of forward bias voltage, the conduction band moves upward and creates a vacancy level in the p-side and as a result tunnelling takes place. The tunnelling current will be maximum at a forward voltage V_p and the corresponding energy band diagram is as shown in Figure 7.2(b).

7.3 VOLT-AMP CHARACTERISTICS OF A TUNNEL DIODE

As forward bias is applied the energy levels on n-side are raised relative to those of the p-side and consequently the electrons in the conduction band on n-side see empty states just across the barrier and the tunnelling takes place. The plot of the voltage versus current characteristics of a tunnel diode is as shown in Figure 7.3. The peak value of current (I_p) will reach at a forward voltage (V_p) where the entire filled state of n-side can see the full

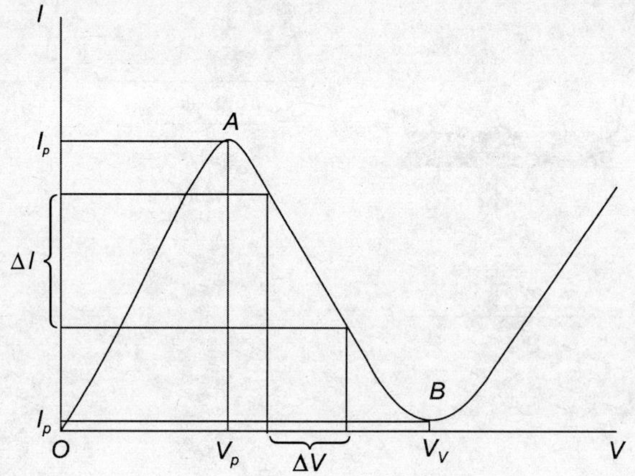

Figure 7.3 Voltage current characteristics of a tunnel diode.

empty state of p-side as per the Figure 7.2(b). As the forward bias is further increased, the energy level on n-side are raised so high that only part of the electron in the conduction band can see available energy level across the barrier [Figure 7.2(c)] and thus tunnelling current will start to decrease from point *A*, which is as shown in Figure 7.3. From V_p to V_v, the voltage is increasing and current is decreasing and the slope over the region *AB* is negative. When the forward bias voltage V increases above the valley voltage V_v, the ordinary injection current I at the p-n junction starts to flow. This injection current increases exponentially in its normal course with the increase of forward biasing voltage. The values of peak current I_p and the valley current I_v determine the magnitude of the negative resistance. For this reason their ratio I_p/I_v is used as a figure of merit for tunnel diode.

An empirical form for the *I-V* characteristic is given by:

$$I = I_p\left(\frac{V}{V_p}\right)\exp\left(1-\frac{V}{V_p}\right) + I_0 \exp\left(\frac{qV}{KT}\right) \tag{7.9}$$

where the first term of the above expression gives the tunnel diode current and the second part gives normal thermal current of a p-n junction diode. The negative differential resistance can be obtained from the derivative of the first term.

$$R = \left(\frac{dI}{dV}\right)^{-1} = -\left[\left(\frac{V}{V_p} - 1\right)\frac{I_p}{V_p} \cdot \exp\left(1 - \frac{V}{V_p}\right)\right]^{-1} \tag{7.10}$$

Figure 7.4 shows a comparative study of the *V-I* characteristics of Ge, GaSb and GaAs tunnel diodes at room temperature [2]. The current ratios of I_p/I_v are 8 : 1 for Ge and GaSb; 12 : 1 for GaSb and GaAs. Again GaSb tunnel diode has the larger negative differential

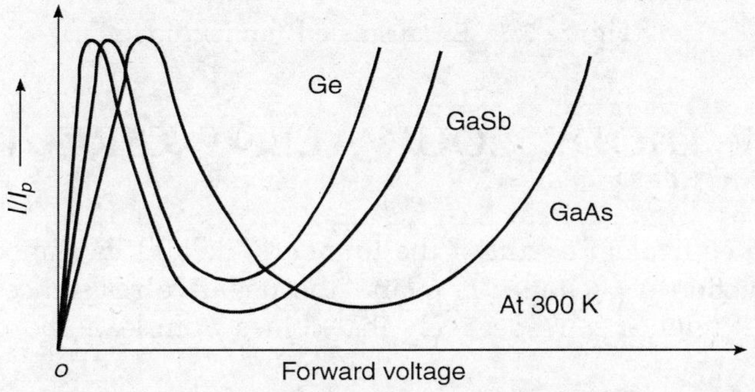

Figure 7.4 Typical *V-I* characteristics of Ge, GaAs and GaSb tunnel diodes at room temperature.

resistance than the other because of its smaller effective mass and small bandgap (0.72 eV). If a reverse bias voltage is applied the height of the barrier is increased above the open circuit value V_0. Thus there are some energy states in the valence band of the p-side, which lie at the same level as allowed empty states in the conduction band of the n-side. Hence these electrons will tunnel through from the p-side to the n-side giving rise to reverse diode current. It may be seen from the diagram that as the reverse bias is increased the diode current increases. Hence the tunnel diode behaves as a good conductor when reverse biased.

7.4 MATERIAL AND CONSTRUCTION OF TUNNEL DIODE

Although tunnel diode can be made from any semiconductor material, initially germanium and gallium antimonide and gallium arsenide have been preferred in practice. All have small forbidden energy gaps and high ion mobility, which is the characteristic leading to good high frequency or high-speed operation. These materials are preferable to silicon and other semiconductors in this regard. A very small tin dot (50 μm dia) is soldered or alloyed to a heavily doped pellet of n-type Ge, GaSb or GaAs. The pellet is then soldered to a Kovar pedestal, used for heat dissipation, which forms the anode contact. The cathode contact is also Kovar, being connected to the tin dot via a mesh screen used to reduce the inductance. The ceramic body is in between the anode and the cathode and it separates the anode from the cathode. Finally the chip is hermetically sealed as shown in Figure 7.5.

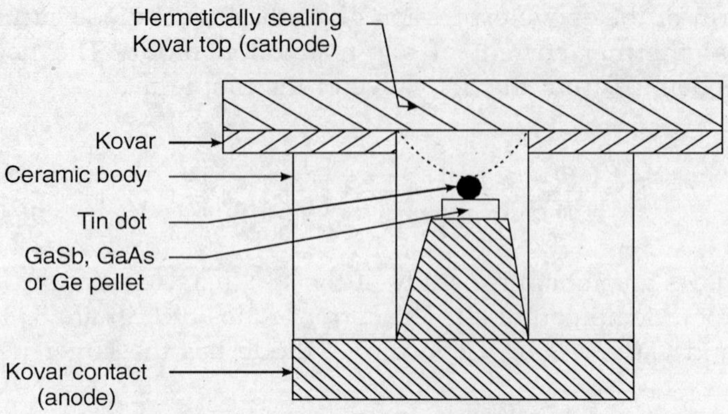

Figure 7.5 Encapsulated tunnel diode.

7.5 TUNNEL DIODE EQUIVALENT CIRCUIT OF OSCILLATOR

Figure 7.6 shows an equivalent circuit of the tunnel diode and its symbol. In the equivalent diode circuit its junction capacitance is C and the negative resistance is R. R_s and L_s are series resistance and inductance respectively that comes from lead and contact.

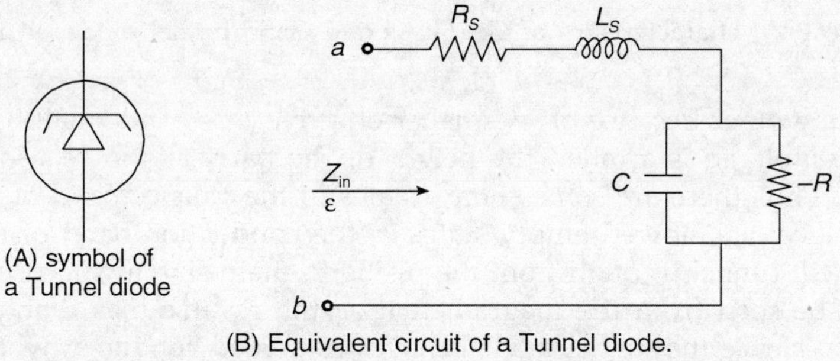

Figure 7.6 Tunnel diode in oscillation mode.

The input impedance Z_{in} of the equivalent circuit across the terminal a and b is given by

$$Z_{\text{in}} = R_s + jwL_s + \frac{R[j/\omega C]}{-R-[j/\omega C]}$$

$$= \left[R_s - \frac{R}{1+(\omega RC)^2} \right] + j\left[\omega L_s - \frac{\omega R^2 C}{1+(\omega RC)^2} \right] \tag{7.11}$$

Now the resistive part of the input impedance (Z_{in}) will be zero at a certain frequency and the reactive part of the impedance will be zero at a second frequency. These two frequencies are denoted by resistive cut-off frequency (f_r) and reactive cut-off frequency (f_x) respectively and are given by:

$$f_r = \frac{1}{2\pi RC}\sqrt{\frac{R}{R_s} - 1} \tag{7.12}$$

$$f_x = \frac{1}{2\pi}\sqrt{\frac{1}{L_sC} - \frac{1}{(RC)^2}} \tag{7.13}$$

Problem 7.1 A tunnel diode has the following parameters:

Negative resistance = 26 Ω
Series resistance = 1 Ω
Junction capacitance = 5 pF
Series inductance = 1 mH
Calculate the resistive and reactive cut-off frequencies.

Solution: From Equation (7.12) the resistive cut-off frequency

$$f_r = \frac{1}{2\pi RC}\sqrt{\frac{R}{R_s} - 1}$$

$$= \frac{1}{2 \times 3.14 \times 26 \times 5 \times 10^{-12}}\sqrt{\frac{26}{1} - 1}$$

$$= 6.12 \text{ GHz}$$

Again from Equation (7.13) the reactive cut-off frequency

$$f_x = \frac{1}{2\pi}\sqrt{\frac{1}{L_sC} - \frac{1}{(RC)^2}}$$

$$= \frac{1}{2 \times 3.14}\sqrt{\frac{1}{1 \times 10^{-3} \times 5 \times 10^{-12}} - \frac{1}{(26 \times 5 \times 10^{-12})^2}}$$

$$= 2.0 \text{ GHz.}$$

7.6 TUNNEL DIODE OSCILLATOR USING CAVITY

The circuit of a tunnel diode oscillator is as shown in Figure 7.7. The circuit is of low cost and of lower power handling capability. Here the tunnel diode is connected across an inductance of suitable value. The oscillator works at the resonant frequency of the junction capacitance and the external inductance. When the negative resistance of the device overcomes the positive resistance of the device, then the oscillation starts. This operation is quite unstable as C_j varies with the temperature and bias voltage, i.e. the undamped resonant frequency varies. By using high Q cavity coupling can be avoided. For the influence of diode weak coupling can minimize capacitance at the operating frequency. The tunnel diode is coupled to the cavity by a probe and a tuning plunger does mechanical tuning. To avoid the effect of inductance loop coupling is avoided.

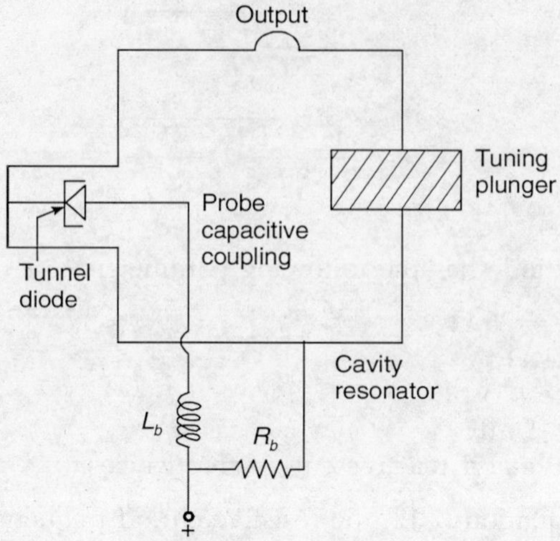

Figure 7.7 Tunnel diode oscillator using cavity.

The major advantages of Tunnel diode oscillator are

I. Circuit is of low cost, simple and can generate moderate amount of power
II. Wide mechanical or electrical tuning range
III. Low noise figure.

These days tunnel diode oscillators can work up to a frequency of 100 GHz and can generate a power of 700 W with a tuning range of 1–4.3 GHz and having the stability of 1 in 10^5.

7.7 TUNNEL DIODE AMPLIFIER WITH CIRCULATOR

Figure 7.8 shows the circuit of a tunnel diode amplifier in a circulator. The circulator is used to separate the input from the output. A tunnel diode amplifier is a medium noise

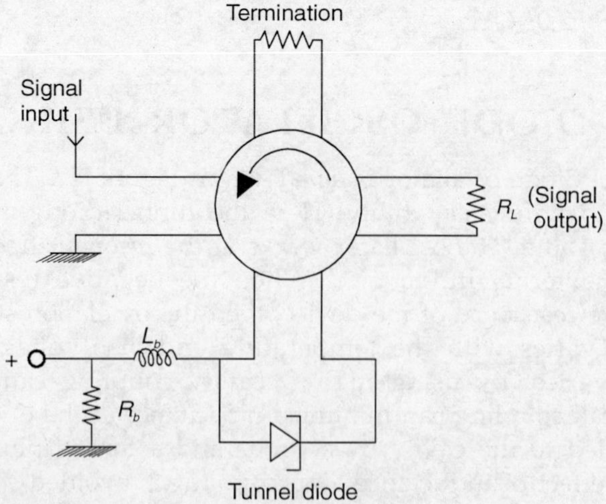

Figure 7.8 Tunnel diode amplifier with circulator.

amplifier but its low power handling capacity limits its use. If the circulator is perfect and has positive real characteristic impedance R_0, an amplifier with infinite gain can be built by selecting a negative resistance tunnel diode whose input impedance has a real part equal to $-R_0$ and an imaginary part equal to zero. The reflection coefficient is infinite. In a tunnel diode amplifier, the resistive load may be connected either in series or in parallel with the diode as shown in Figure 7.9(a) and (b) respectively.

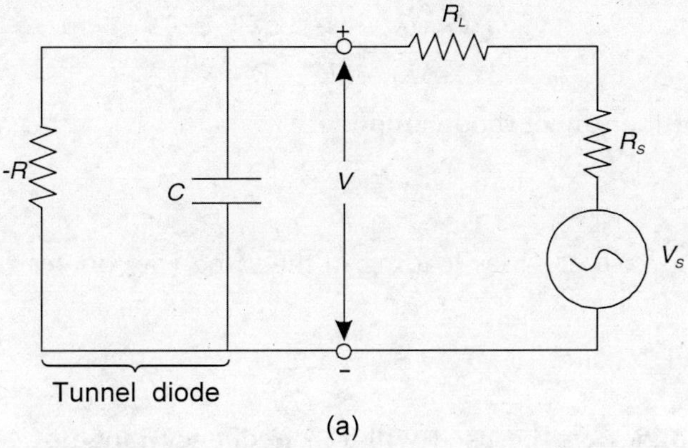

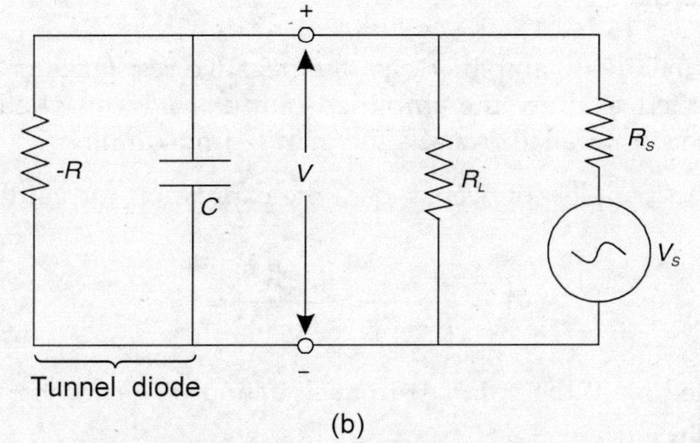

Figure 7.9 (a) Load is connected in series with the diode and (b) load is connected in parallel with the diode.

In case of parallel conduction of load, as V is the voltage appearing across the load, power dissipated across the load

$$P_L = \frac{V^2}{R_L} \tag{7.14}$$

One part of the load power is the amplified power of the tunnel diode and if A be the gain of the tunnel diode amplifier

$$P_{\text{in}} = \frac{V^2}{AR_L} \tag{7.15}$$

Other part of the load power is generated by the negative resistance (R) of the device and is given by:

$$P = \frac{V^2}{R} \tag{7.16}$$

Thus combining Equations (7.14) and (7.15) and Equation (7.16), we get

$$\frac{V^2}{AR_L} + \frac{V^2}{R} = \frac{V^2}{R_L}$$

and thus the gain of the tunnel diode amplifier

$$A = \frac{R}{R - R_L} \tag{7.17}$$

On the other hand in case of series loading in the same way one can calculate the gain of the amplifier as:

$$A = \frac{R_L}{R_L - R} \tag{7.18}$$

A tunnel diode amplifier can be used within a frequency range of 4–15 GHz and with a maximum gain of 15 dB.

Problem 7. 2 A tunnel diode amplifier has the negative resistance of 26 Ω. Now a load of 24 Ω is to be connected to draw the amplified output. Suggest whether the load is to be connected in series or in parallel to make the gain (A) maximum.

Solution: For a parallel connected load, the expression for gain from Equation (7.17) is

$$A = \frac{R}{R - R_L} = \frac{26}{26 - 24} = 13$$

For a series connected load, the gain (A) from Equation (7.18) can be written as

$$A = \frac{R_L}{R_L - R} = \frac{24}{24 - 26} = -12$$

Therefore,

$$|A| = 12$$

Thus the gain will be more when the load is connected in parallel.

7.8 TUNNEL DIODE SWITCH

Tunnel diode can be used as a high-speed switch. Since tunnelling takes place at the speed of light, the transient response is limited only by total shunt capacitance, and peak driving current switching time is of the order of a nanosecond but minimum switching time is as low as 50 pico second.

7.9 RESONANT TUNNELLING DIODE (RTD)

This is another type of tunnel diode, whose operation is based on resonant tunnelling effect. RTD type of device have a semiconductor double-barrier structure that comprises of several heterojunctions and one quantum well be in the conduction band. In the year 1987 Kesan [3] invented and developed Quantum Well Injection Transit Time (QWITT) diode, whose principle is based on resonant tunnelling through a quantum well between two Al_xGa_{1-x} As barriers, in series with a drift region of undoped GaAs and this structure is then placed between two n^+–GaAs regions to form contacts as shown in Figure 7.10(a). The conduction band diagram of the device of equilibrium is shown in Figure 7.10(b).

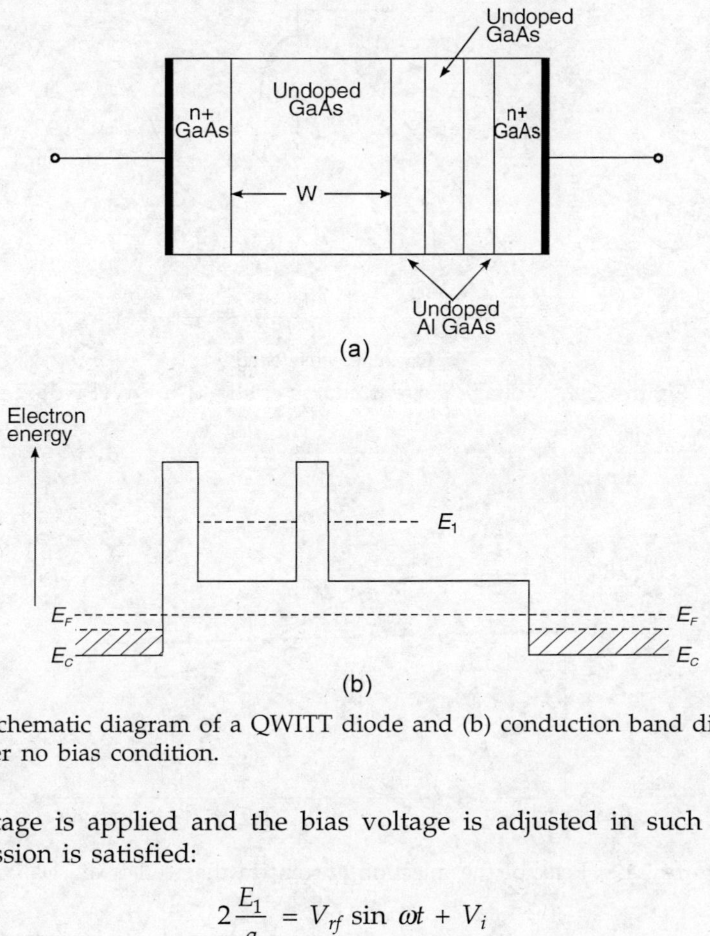

Figure 7.10 (a) Schematic diagram of a QWITT diode and (b) conduction band diagram of QWITT under no bias condition.

The *rf* voltage is applied and the bias voltage is adjusted in such a way that the following expression is satisfied:

$$2\frac{E_1}{q} = V_{rf} \sin \omega t + V_i \tag{7.19}$$

where V_i is the effective input voltage across the well and E_1 is the energy level in the well. Figures 7.10(a) and (b) show the required condition for tunnelling. By proper adjustment of biasing voltage the Fermi level in the n^+–GaAs is brought below E_1 and then the *rf* signal is superimposed on dc biasing voltage. The tunnelling effect will now start and the peak value of tunnelling current is obtained at $\omega t = \pi/2$. Also the electrons will now traverse through the drift region at their saturation velocity resulting in a dynamic negative resistance. The same way the diode can be biased in the negative half of the *rf* cycle (Figure 7.11).

The QWITT diode is less efficient than other transit time devices, but have the advantages like low noise and can operate at a very high frequency. Depending on the bias level, the device may permit injection of carriers into the drift region at more favourable phase angles than other transit time devices. So higher efficiencies can be expected. The characteristics of the injection region and therefore the injected current pulse can be adjusted by changing the physical dimension of the quantum well structure as seen in the Figure 7.12.

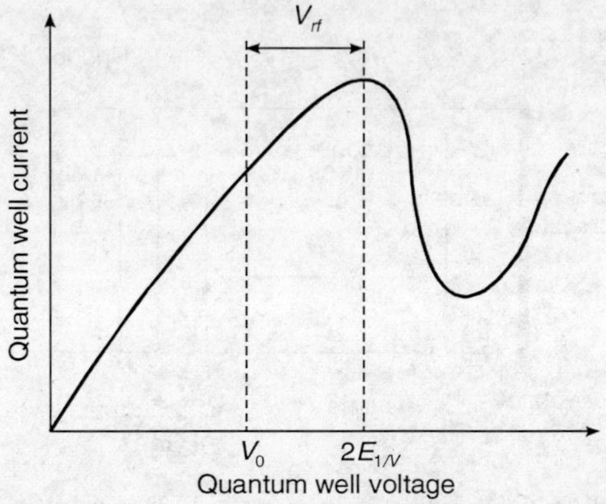

Figure 7.11 Voltage current characteristic of a QWITT diode.

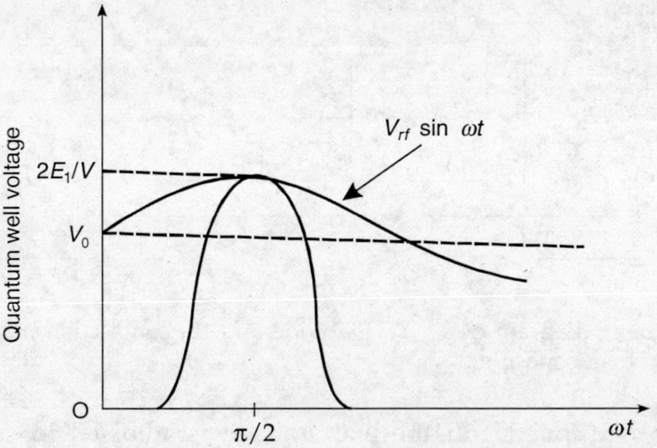

Figure 7.12 Peak of the injection current in the well is at $\omega t = \pi/2$.

REFERENCES

1. Esaki, L., New phenomenon in narrow Ge p-n junction, *Phys. Rev.*, Vol. 109, p. 603, 1958.

2. Tunnel Diode Manual, **TD-30**, Radio Corporation of America, Semiconductor and materials division, Somerville, N.J., 1963.

Tunnel Diode Manual, General Electric Company, Semiconductor Products Dept., Liverpool, N.Y., 1961.

3. Kesan, V.P., A new transit time device using quantum–well injection, *IEEE Electron Device Letters*, **EDL-8**, No. 4, 129, 1987.

SUGGESTED READINGS

1. Sze, S.M., *Semiconductor Devices: Physics and Technology*, 2nd ed., John Wiley & Sons, 2002.

2. Liao, S.Y., *Semiconductor Electronic Devices*, Prentice Hall Inc., Englewood Cliffs, N.J., 1990.

3. Millman, J. and Halkias, C.C., *Electronic Devices and Circuits*, Tata McGraw-Hill Edition, 1996.

4. Liao, S.Y., *Microwave Devices and Circuits*, 3rd erd., Prentice-Hall of India, New Delhi, 1995.

PROBLEMS

1. Write down the condition for tunnelling in a Tunnel diode.
2. Why tunnelling is not possible in an ordinary p-n junction diode?
3. How a Tunnel diode exhibits negative resistance over a specific voltage range and explains the significance of negative resistance.
4. Why GaSb is the most suitable material for the fabrication of Tunnel diode?
5. Mention some of the applications of a Tunnel diode.
6. A Tunnel diode has the following parameters:

   Negative resistance = 18 Ω
   Series resistance = 2 Ω
   Resistive cut-off frequency = 8 GHz
   Series inductance = 1 mH.

   Calculate (i) the junction capacitance and (ii) the reactive cut-off frequency.
7. A Tunnel diode amplifier has the negative resistance of 20 Ω. Now a load of 12 Ω is connected in series. Calculate the gain (A) of the amplifier.
8. A microwave Tunnel diode has a negative resistance R_n and the resonant circuit has a resistance R_r. Derive an equation for the gain of a microwave Tunnel diode amplifier.
9. Mention the major advantages of a Tunnel diode oscillator.
10. Explain the Resonant tunnelling effect and how on the basis of this effect a QWITT diode works?

Chapter 8

Schottky Barrier Diodes

8.0 INTRODUCTION

Schottky Barrier diodes are metal semiconductor junctions having rectifying properties which are used as detectors and mixers at microwave frequencies. In 1904, Sir. J.C. Bose [1] first showed that a point contact metal semiconductor device, i.e., a metal whisker on a piece of semiconductor could be used as a rectifier at microwave frequencies. Later on metal semiconductor junctions based on new technology of depositing thin films of metal on semiconductors at very high vacuum have been extensively studied and are used as microwave detectors, mixers, modulators and demodulators. Schottky [2], Mott [3], and Bethe [4] developed the theory of metal semiconductor contacts. Rhoderick and Willams (1988) [5] have given a very detailed theoretical and experimental review of metal-semiconductor contacts.

8.1 THEORY

The work function of a metal plays a vital role in determining the properties of a metal semiconductor junction. The energy $e\phi_m$ required for removal of an electron in metal from its Fermi level to the vacuum outside is called the work function of the metal. Similarly the energy $e\phi_s$ denoting the energy difference of an electron in semiconductor from its Fermi level to the vacuum outside is called the work function of the semiconductor. The energy required to lift an electron from the lowest level of conduction band to vacuum level is called electron affinity $e\chi$.

Consider now a metal with work function $e\phi_m$ and a semiconductor of work function $e\phi_s$ to contact and form a metal semiconductor junction. Electron transfer will take place until the two Fermi levels are brought to the same level.

If the work function of the metal is greater than that of the semiconductor, then the metal Fermi level is lower than the Fermi level of the semi-conductor before the contact is made (Figure 8.1). After the contact, the electron energy, in the semiconductor must be lowered relative to the electron energy in the metal. The electrons, leaving the semi-conductor leave behind an unbalanced positive charge arising out of positive donor ions in the depletion region and produce a thin sheet of negative charge on metal as shown in Figure 8.2. The resulting electric field causes a band bending near the surface. Figure 8.3 shows the vacuum level, band bending and the depletion region for a metal semiconductor

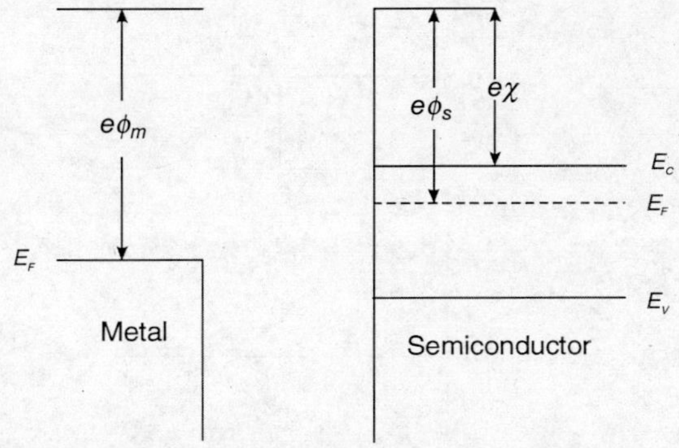

Figure 8.1 Metal-semiconductor band diagrams before contact $e\phi_m$ = Metal work function, $e\phi_s$ = Semiconductor work function, $e\chi$ = Electron affinity.

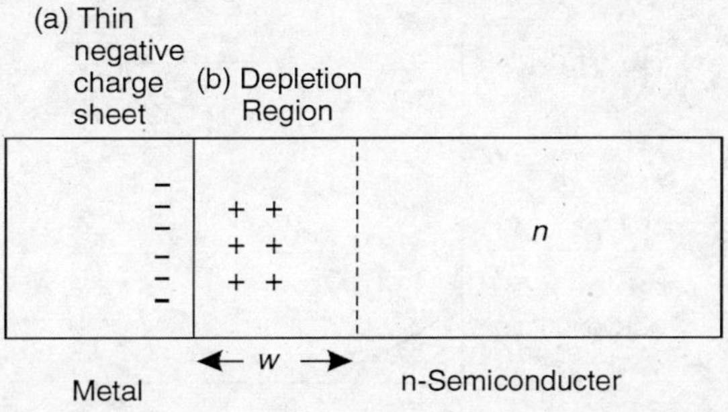

Figure 8.2 Metal-semiconductor junction after contact with (a) thin sheet of negative charge on metal side of junction and (b) depletion region of width w on the semiconductor side consisting of immobile positive donor ions.

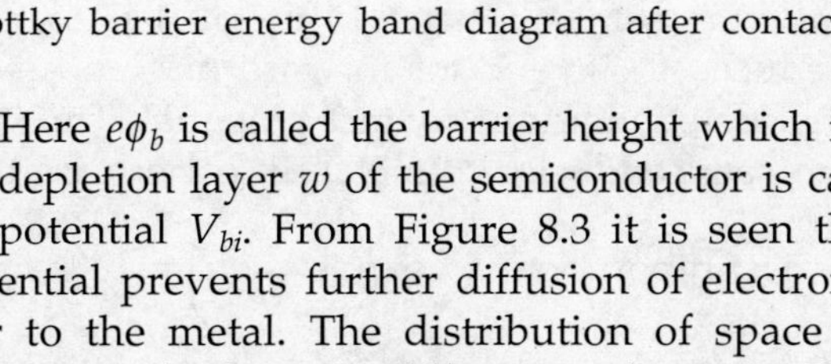

Figure 8.3 Schottky barrier energy band diagram after contact at equilibrium.

junction at equilibrium. Here $e\phi_b$ is called the barrier height which is equal to $e\phi_m - e\chi$. The voltage drop across the depletion layer w of the semiconductor is called the built-in-voltage or equilibrium contact potential V_{bi}. From Figure 8.3 it is seen that $V_{bi} = e\phi_m - e\phi_s$. The equilibrium contact potential prevents further diffusion of electrons from the conduction band of semiconductor to the metal. The distribution of space charge, semiconductor depletion of a Schottky Barrier diode in equilibrium is shown in Figure 8.4.

The current flowing through a metal semiconductor junction is caused by (a) I_0 due to electron emission from metal to semiconductor junction over the barrier $(\phi_m - \chi)$ and (b) due to electron flow from semiconductor conduction band over the contact potential into the metal.

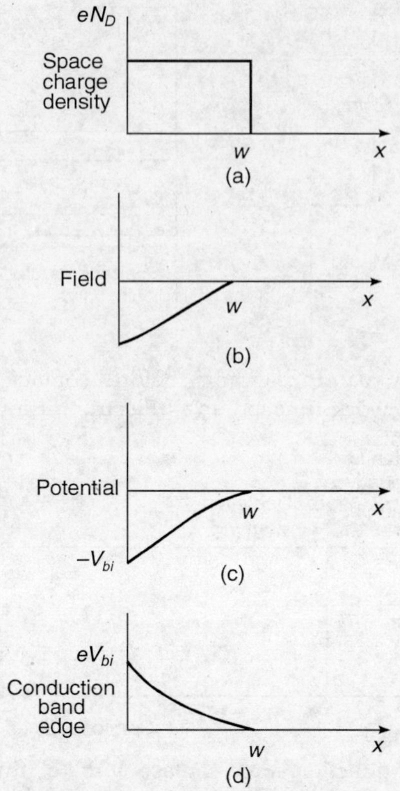

Figure 8.4 Distribution of space charge, electric field, potential and conduction band edge in the depletion layer of a Schottky barrier diode.

It can be shown that the probability of electron emission from metal surmounting the barrier ϕ_b is proportional to $\exp\left(-\frac{e\phi_b}{RT}\right)$. Thus the current I_0 from metal to semiconductor is proportional to $\exp\left(-\frac{e\phi_b}{RT}\right)$ for forward as well as reverse bias.

For flow of current from semiconductor to metal it is to be noted that the contact potential difference is reduced to $V_{bi} - V$ for forward bias and is increased to $V_{bi} + V$ for reverse bias, where V_{bi} is the equilibrium contact potential. Thus for forward bias there is larger electron flow from semiconductor to metal. In either case the flow of electron from metal to semiconductor is proportional to $\exp\left(-\frac{e\phi_b}{RT}\right)$ which is unaffected by the bias voltage. The diode equation for the metal-semiconductor junction is thus

$$I = I_0\left(\exp\frac{eV}{kT} - 1\right)$$

$$= A\exp\left(\frac{-e\phi_b}{kT}\right)\left(\exp\frac{-eV}{kT} - 1\right)$$

and is similar to the diode equation for p-n junction. In this case the reverse saturation current I_0 depends on electron flow from metal to semiconductor over the barrier which is unaffected by the bias voltage. The reverse saturation current depends on the height of the barrier ϕ_b for electron injection from metal to semiconductor.

The effect of the forward and reverse bias on the energy band diagram and a typical current-voltage characteristic is shown in Figure 8.5. The diode equation also holds good for a metal (p-type) semiconductor junction. When a positive voltage is applied to the semiconductor side (p-type) of the junction, forward current passes as holes flow from the p-semiconductor to metal since potential barrier is lowered. With reverse biasing the potential barrier increases and the current flow due to holes become very small.

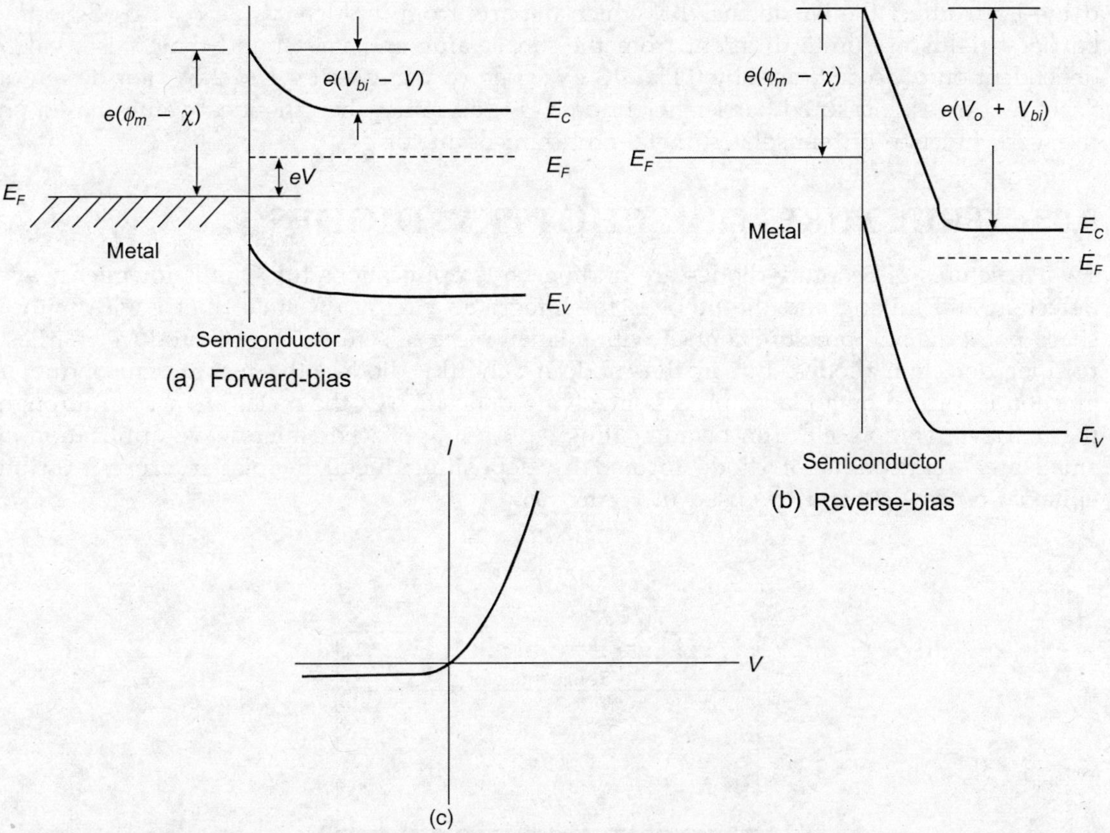

Figure 8.5 The effect of forward and reverse bias or energy band diagram and a typical current voltage characteristic of a Schottky Barrier diodes.

In Schottky Barrier diodes the forward current flows due to majority carriers from semiconductor to metal and there is no minority carrier current injection. Hence the phenomena of storage and recombination of minority carriers along with associated time delays do not occur in rectifying Schottky Barrier diodes unlike p-n junction rectifiers. Thus very fast switching can take place and the devices are very suitable for detection of high frequency signals.

8.2 SURFACE STATES IN PRACTICAL SCHOTTKY BARRIER DIODES

In practice Schottky Barrier Diodes involve a termination of crystal semiconductor lattice at the metal semiconductor boundary. The termination leads to production of surface states due to incomplete covalent bonds. These surface states change the barrier height in real case from the ideal metal-semiconductor barrier height. Also a thin oxide layer is always present at the semiconductor surface and there is no atomically sharp discontinuity between the metal and the semiconductor at the junction. Thus silicon crystals are covered by a 10–20A° of thin SiO_2 layer, even when cleaved or etched in clean condition although electrons can easily tunnel through this thin layer. Both the surface states and the thin oxide layer affect the barrier height, which departs from its ideal value $\phi_m - \phi_s$. Schottky barrier heights are quite different from the ideal value and it is found that ϕ_b is weakly dependent on ϕ_m. ϕ_b increases by 0.1 to 0.3 eV when ϕ_m increases by 1 to 2 eV. For design of Schottky diodes, measured barrier height data is used rather than the ideal values based on the work function of the isolated metal and semiconductor.

8.3 STRUCTURE OF SCHOTTKY DIODES

Two structures of Schottky diodes are finding wide applications [6]. One is for microwave detection and mixing and the other is for integrated circuits. A small metal wire with a sharp point making pressure contact with a small piece of semiconductor produced earliest rectifier detection diodes. But in the modern Schottky diodes, the metal semiconductor junction is formed by a planar process involving deposition of the metal in vacuum through thermal evaporation, electron beam coating, sputtering etc. For microwave application, a small area semiconductor diode formed by depositing metal by planar process on an epitaxial n or n$^+$ substrate is used in Figure 8.6.

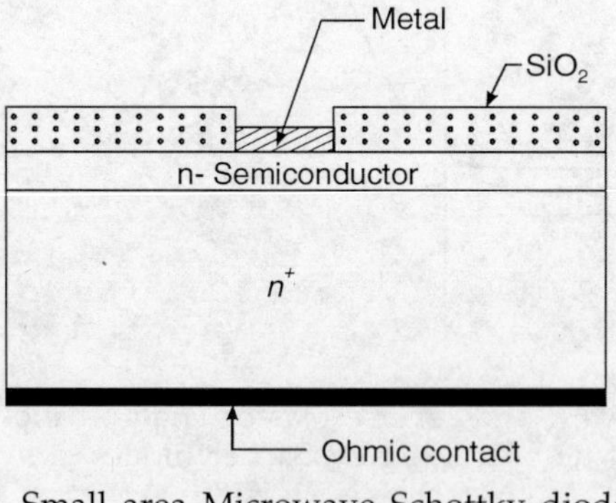

Figure 8.6 Small area Microwave Schottky diode detector.

The series resistance and the diode capacitance have to be minimized for good performance. The cut-off frequency for forward bias is given by

$$f_c = \frac{1}{2\pi R_f C_f}$$

where R_f and C_f are the values for forward bias of 0.1 V. A typical variation of f_c against epitaxial donor concentration of m-n Si and m-n GaAs structures for a junction diameter of 10 μm is shown in Figure 8.7. It is seen that the metal-n GaAs diode has a much higher cut-off frequency than that of metal-n Si diode, which arises because of the much higher electron mobility in GaAs, compared to silicon. Higher junction diameter is found to lower the cut-off frequency due to higher capacitance of the device.

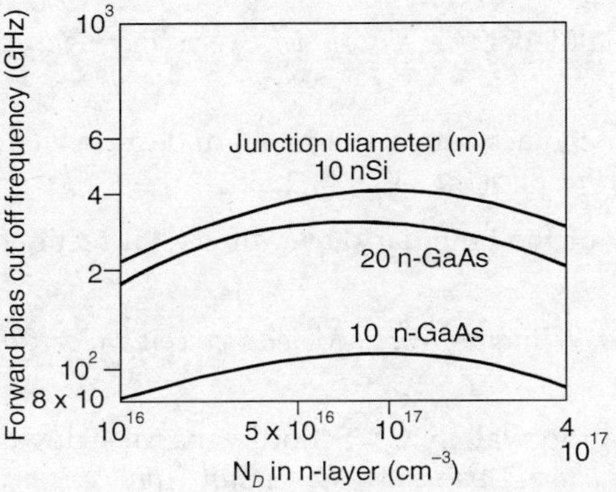

Figure 8.7 Variation of f_{co} with doping concentration in metal-n-GaAs and metal-n-Si junctions.

Another Schottky diode structure particularly suitable for integrated circuit is the metal overlap structure shown in Figure 8.8. This structure gives well forward *I–V* characteristics and low leakage current. The structure can be formed as an integral part of metallization in integrated circuits.

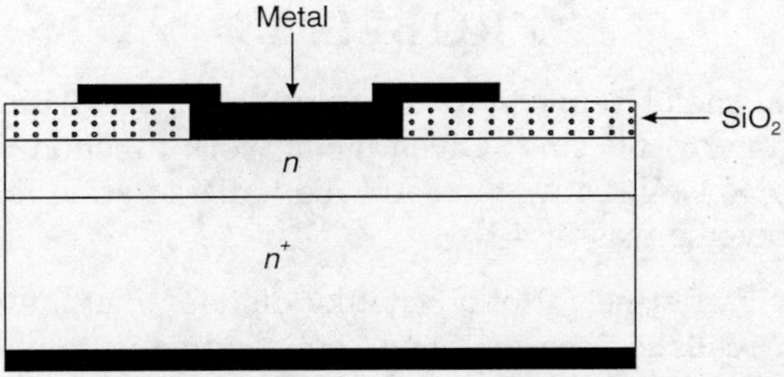

Figure 8.8 Metal overlap Schottky structure.

A metal semiconductor junction will behave as an ohmic contact when the semiconductor is highly doped, i.e., metal-n^+ -n or metal-p^+ -p. The heavily doped surface can be formed by shallow diffusion or alloy regrowth. The heavily doped n^+ or p^+ layer in the metal semiconductor junction leads to extremely thin depletion layer. The carriers can tunnel easily through this thin depletion layer from semiconductor to metal and the diffusion of the carriers over the contact potential plays a minor role. Thus very heavy current flows through tunnelling of the electrons from the semiconductor to the metal irrespective of the polarity of the bias voltage. Thus current can flow in both the directions making the metal semiconductor junction behave like an ohmic contact.

REFERENCES

1. Bose, J.C., U.S. Patent 775840 (1904).
2. Schottky, W. and Halbleiter A., *Theory der Sperrschict, Naturissensch aften*, Vol. 26, 843, 1938.
3. Mott, N.F., Note on contact between a metal and an insulator or semiconductor, *Proc. Cambridge Philosophical Society*, **34**, 568, 1938.
4. Bethe, H.A., Theory of the boundary layer of crystal rectifiers, *MIT Radiation Lab*, **Rep.** 43-52, 1942.
5. Rhoderick, E.H. and Williams, R.H., *Metal Semiconductor Contacts*, 2nd ed., Clarenden Press, Oxford, 1988.
6. Irvine, J.C. and Vanderwal, N.C., Schottky barrier devices in H.A. Watson (Ed.) *Microwave Semiconductor Devices and the Circuit Applications*, McGraw-Hill, New York, 1968.

SUGGESTED READINGS

1. Sze, S.M., *Physics of Semiconductor Devices*, 2nd ed., Wiley Eastern, 1983.
2. Streetman, B.G. and Bannerjee, S., *Solid State Electronic Devices*, 5th ed., Prentice-Hall of India, New Delhi, 2002.

PROBLEMS

1. Consider an ideal metal semiconductor Schottky diode with n-type silicon having donor concentration of 10^{17} cm^{-3}. Given metal work function ϕ_m = 4.5 eV and silicon electron affinity χ = 4 eV. Draw the band diagrams to scale for equilibrium, forward bias of 2 V and reverse bias of 4 V.
2. What is meant by barrier height of a Schottky diode? What are the factors which affect barrier height in realistic case of metal semiconductor contact? Why the speed of response of Schottky diode rectifier is much greater than the p-n junction rectifier?

Chapter 9

Microwave Bipolar Transistors

9.0 INTRODUCTION

The transistor is a multi-junction semiconductor device, where both the types of charge carriers take part in current-carrying mechanism. William Shockley and his co-workers at Bell Laboratory had invented the bipolar transistor in 1947. The two types of Bipolar Junction Transistors (BJT) are n-p-n and p-n-p, when a p-type substrate forms the transistor, an n-type region is thermally diffused through an oxide window into the p-type substrate. A heavily doped p^+ region is then diffused into the n-type region. Finally metal contacts are made with the three layers. The n-p-n bipolar transistor is the complimentary structure of the p-n-p bipolar transistor.

Microwave power transistor technology have been advancing significantly during the past three decades. The microwave transistor is a nonlinear device and its principle of operation is similar to that of the low frequency device. Higher interelectrode capacitance, lead inductance and transit time limits the operation of low frequency transistor in the high frequency range. Thus proper geometry to reduce the above effects, process control, heat sinking and packaging are first and foremost criteria for the designing of a high frequency transistor. All microwave bipolar transistors are planar in form and is of n-p-n type. As mobility of electron (μ_e) is greater than that of hole (μ_h), hence electron emitted by the emitter of an n-p-n transistor will reach collector faster than the hole and which in turn increases the speed of operation. These days microwave Bipolar Junction Transistors are capable of generating power up to a frequency of 22 GHz. The majority of Bipolar Transistors of current interest are fabricated from silicon, although GaAs devices offer prospects for improvements in operating frequency, in high temperature and in radiation hardness. On the other hand the silicon BJTs are of low cost, more reliable integrative and offers higher gain. It has moderate noise figure when used as a microwave amplifier and $1/f$ (Flicker noise) noise characteristics. For these reasons, the silicon bipolar transistors dominate in amplifier applications for the lower microwave frequencies and are often the devices of choice for local oscillators.

9.1 PHYSICAL STRUCTURE

Because of high frequency operation the geometry of low frequency BJTs are modified by Sobol and Sterzer [1], which are as shown in Figure 9.1(a), (b) and (c). The inter digitated

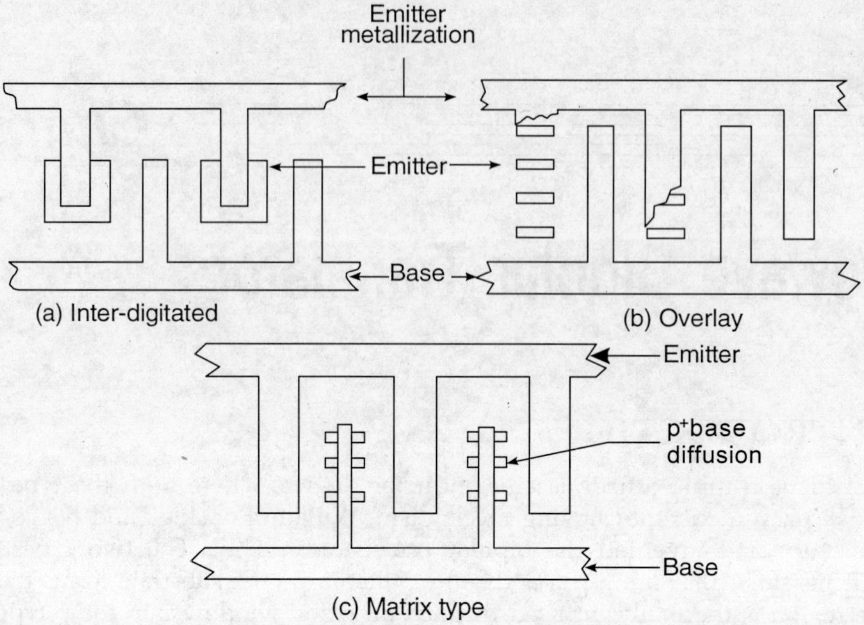

Figure 9.1 Geometrical structures of microwave power transistor.

geometry basically consists of large number of emitter strips alternating with base strips. Both of these are metallized. The overlay geometry has a large number of segmented emitters overlaid through a number of wide metal strips. The matrix or mesh geometry has emitter that forms the grid, the base filling the meshes of this grid with a p^+ contact area in the middle of each mesh. Inter-digitated structure is suitable for small signal applications in the L, S and C bands whereas overlay and matrix structures are useful as power devices in the VHF and UHF regions.

9.2 PRINCIPLE OF OPERATION

The bipolar junction transistors are an active three terminal device, which is commonly used as an amplifier or switch. A BJT can operate in four different modes:

(i) **Normal mode:** In this mode emitter junction of n-p-n transistor is forward biased and collector is reverse biased. Generally at ON state a transistor remains in the normal mode.

(ii) **Saturation mode:** When both the junctions are forward biased, the transistor is in its saturation mode with very low resistance and acts like a short circuit.

(iii) **Cut-off mode:** If both transistor junctions are reverse biased, the transistor is operated in cut-off mode and the transistor acts like an open circuit. Thus saturation and cut-off modes are equivalent to the ON and OFF state of a switch.

(iv) **Inverse mode:** A transistor is said to be in inverse mode when the emitter is reverse biased and collector is forward biased. In practice transistor is not commonly used in inverse mode. The microwave signal makes the base emitter junction forward biased at the negative portion of the signal and as a result electrons from the emitter region diffuses into the base and reaches the collector (n-layer) by drift mechanism. In the other half of the input signal the transistor remains in the OFF state. Thus a pulse of current flows through the load connected in the collector circuit.

9.3 PERFORMANCE PARAMETER

In high frequency operation the performance of a microwave transistor depends on the cut-off frequency (f_c) and maximum frequency of oscillation (f_{max}) rather than the two current gains α and β.

Now the cut-off frequency depends on the delay of the carrier results due to their movement from emitter to collector,

thus
$$f_c = \frac{1}{2\pi\tau_{ec}} \tag{9.1}$$

where
$$\tau_{ec} = \tau_e + \tau_b + \tau_d + \tau_c \tag{9.2}$$

and τ_e = Emitter base junction transit time

τ_b = Base transit time

τ_d = Collector depletion layer transit time

τ_c = Collector depletion layer charging time.

Maximum frequency of operation is higher than f_c because although β falls to unity at this frequency, power gain does not.

Therefore
$$f_{max} = \sqrt{\frac{f_c}{8\pi r_b' C_c}} \tag{9.3}$$

where

r'_b = Base resistance

C_c = Collector capacitance

9.4 POWER FREQUENCY LIMITATIONS

As the frequency is increased the output power drops mainly due to low impedance of the junction capacitance. As per Johnson [2] the power frequency limitation is due to the

(i) maximum velocity of charge carriers in a semiconductor and which is 6×10^6 cm/s in Ge or Si.

(ii) the maximum electric field that can be applied to a semiconductor is 2×10^5 V/cm.

(iii) the base width determines the maximum current.

Let V_m be the maximum applied voltage and the corresponding current be I_m, and if the charge carrier takes τ_{ec} time to cross the length L of the device

Then,

$$V_m \cdot f_c = \frac{V_m}{L} \cdot \frac{L}{2\pi\tau_{ec}}$$

$$= \frac{E_m v_s}{2\pi}. \tag{9.4}$$

where

$$E_m = \text{Maximum electric field}$$

$$= \frac{V_m}{L}$$

and

$$v_s = \text{Saturated drift velocity of the carrier} = L/\tau_{ec}$$

Again if C_{bc} be the collector base capacitor of the transistor then its corresponding reactance

$$X_c = \frac{1}{\omega_c C_{bc}} = \frac{1}{2\pi f_c C_{bc}} \tag{9.5}$$

From Equation (9.4) we can write

$$(I_m X_c) \cdot f_c = \frac{E_m v_s}{2\pi} \tag{9.6}$$

Now maximum power dissipated across X_c is

$$P_m = I_m^2 X_c$$

or

$$I_m = \left(\frac{P_m}{X_c}\right)^{1/2} \tag{9.7}$$

Substituting the value of I_m from Equation (9.7) to Equation (9.6) we get

$$(P_m X_c)^{1/2} f_c = \frac{E_m v_s}{2\pi}$$

or

$$f_c = \frac{E_m v_s}{2\pi} \cdot \frac{1}{X_c^{1/2} \cdot P_m^{1/2}} \tag{9.8}$$

From Equation (9.8) it is clear that keeping v_s, E_m and $X_c^{1/2}$ constant, the frequency

$$f_c \propto \frac{1}{\sqrt{P_m}}$$

Thus for a microwave BJT the output power decreases with the increase of frequency. For a given value of $E_m v_s$ the maximum power that can be delivered to the carriers through the transistor is traversing till infinite if the cross-section of the transistor is made very large.

The gain frequency relation is given by

$$(G_m \,.\, V_{th}\ V_m)^{1/2} \,.\, f = \frac{E_m v_s}{2\pi} \tag{9.9}$$

where

G_m = maximum available power gain

V_{th} = kT/e is the thermal voltage

k = Boltzmann constant

f = Frequency of operation

The maximum available power gain of a transistor is derived by Johnson [2]

$$G_m = \left(\frac{f_c}{f}\right) . \frac{Z_{\text{out}}}{Z_{\text{in}}} \tag{9.10}$$

where Z_{out} and Z_{in} are output and input impedances respectively. Considering the electrode series resistance to be zero, the ratio of output impedance to the input impedance can be written as

$$\frac{Z_{\text{out}}}{Z_{\text{in}}} = \frac{C_{\text{in}}}{C_{bc}} \tag{9.11}$$

where C_{in} is the input capacitance and C_{bc} is the output capacitance arising between base and collector. If the maximum total carrier charge Q_m moves to the collector in a carrier base transit time τ_b with a thermal voltage V_{th}, the input capacitance (C_{in}) and the emitter diffusion capacitance C_d are related by

$$C_{\text{in}} \approx C_d \cong \frac{Q_m}{V_{th}} \cong \frac{I_m \cdot \tau_b}{V_{th}} \tag{9.12}$$

The output capacitance is given by

$$C_{\text{out}} = C_{bc} = \frac{I_m \cdot \tau_{ec}}{V_m} \tag{9.13}$$

Thus Equations (9.6), (9.8) and (9.9) also give the limitations of current-frequency, power-frequency and gain-frequency of a microwave bipolar transistor. At present microwave transistor power, frequency and gain limits are:

Output power: 20–150 mW for frequency between 4–10 GHz.

Power gain: 31 ± 1.5 dB over the frequency range of 4–6 GHz with an average power output of 15 mW.

Noise figure: 3.3–14 dB for frequencies between 4–8 GHz.

Problem 9.1 A certain silicon microwave transistor has the following parameters

Reactance $X_c = 1\ \Omega$

Maximum electric field $E_m = 1.6 \times 10^5$ V/cm

Saturated drift velocity $v_s = 4 \times 10^5$ cm/s

Maximum allowable power $P_m = 6.48$ W.

Determine the transit time cut-off frequency.

Solution: From Equation (9.8) we get

$$f_c = \frac{1}{X_c^{1/2} P_m^{1/2}} \cdot \left(\frac{E_m v_s}{2\pi} \right)$$

$$= \frac{1}{1 \times (6.48)^{1/2}} \cdot \left(\frac{1.6 \times 10^5 \times 4 \times 10^5}{2 \times 3.14} \right)$$

$$= 4 \text{ GHz}$$

9.5 APPLICATIONS

Microwave Bipolar Transistor is used in L-band transmitters for telemetry system and phased array radar systems, it is also used in L and S-band transmitters for communication systems

9.6 HETEROJUNCTION BIPOLAR TRANSISTORS (HBTs)

Bipolar transistor may be a homojunction or a heterojunction type. Transistor junction formed by two similar materials such as silicon to silicon is known as homojunction transistor, whereas the transistor junction formed by two different materials such as Ge to GaAs is called a heterojunction transistor.

9.6.1 Structure of heterojunction transistors

Two materials can be used to fabricate a heterojunction transistor when their lattice constants are matched. Otherwise mismatch of lattice constants introduce a large number of interface states and degrade the heterojunction operation. Figure 9.2 shows the model diagram of a heterojunction transistor formed by n-Ge, p-GaAs and n-GaAs materials. The energy band diagram for isolated n-Ge and p-GaAs is as shown in Figure 9.3, where the

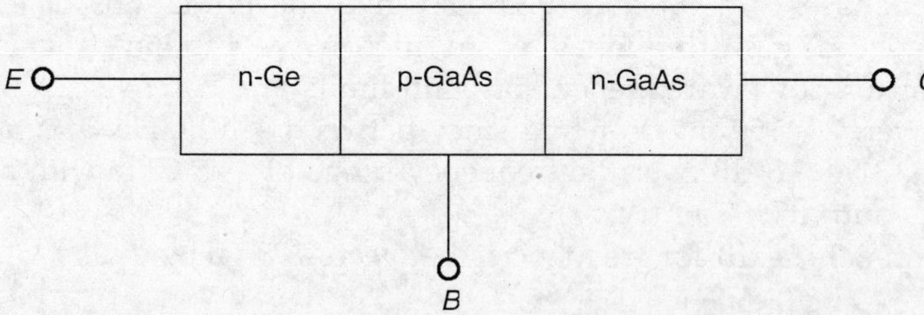

Figure 9.2 Model diagram of a heterojunction transistor.

Fermi energy levels are not aligned.

Here ϕ_1 and ϕ_2 are the work functions of n-Ge and p-GaAs respectively. Let E_g be the bandgap energy in eV and ψ be the electron affinity in eV.

Now from Figure 9.3 we can write

$$\Delta E_c = \psi_1 - \psi_2 \tag{9.14}$$

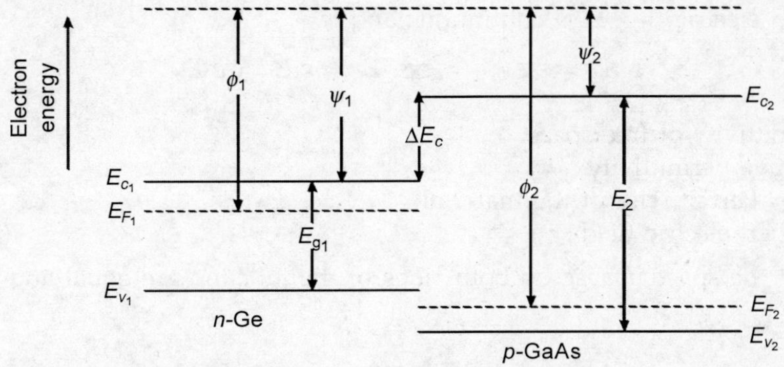

Figure 9.3 Energy band diagram for a heterojunction transistor when layers are isolated.

$$\Delta E_v = E_{g_2} - E_{g_1} - \Delta E_c \tag{9.15}$$

when the two semiconductor materials are joined together, their Fermi energy levels are aligned and their energy bands are depleted at the junction as shown in Figure 9.4.

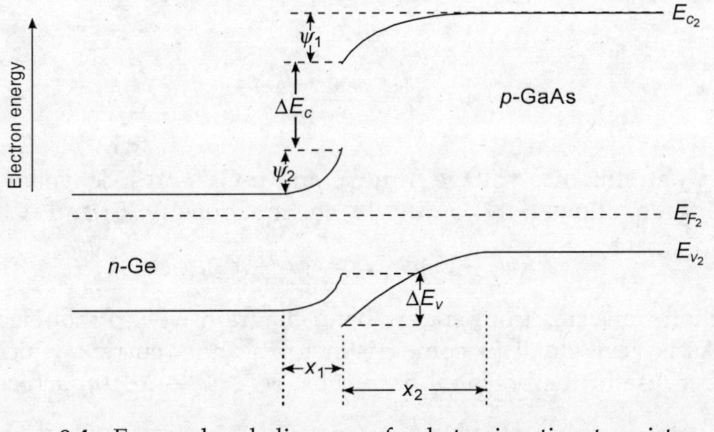

Figure 9.4 Energy band diagram of a heterojunction transistor.

After junction formation the electron in n-Ge are injected into the p-GaAs and the holes are injected from p-GaAs to n-Ge until their Fermi energy levels are aligned and as a result the energy bands at the junction are depleted or bent. Thus in both sides of the junction the bending energy creates a built-in voltage (Figure 9.4). The total built-in voltage is expressed as

$$\psi_0 = \psi_1 + \psi_2 \tag{9.16}$$

where

ψ_1 = Built-in voltage in n-Ge
ψ_2 = Built-in voltage in p-GaAs

At the junction electric flux D is continuous and is

$$D = \varepsilon_0 \varepsilon_{r1} E_1 = \varepsilon_0 \varepsilon_{r2} E_2 = \varepsilon_1 E_1 = \varepsilon_2 E_2 \tag{9.17}$$

where

ε_0 = Permittivity of free space
ε_r = Relative permittivity
$\varepsilon = \varepsilon_0\, \varepsilon_r$ = Permittivity of the material
E = Applied electric field

Assuming that the space charge on both sides of the junction are equal and is given by:

$$x_1 N_d = x_2 N_a \tag{9.18}$$

where

x_1 = Depletion width in n-Ge
x_2 = Depletion width in p-GaAs
N_d = Donor density
N_a = Acceptor density

The electric field on both sides of the junction can be written as

$$E_1 = \frac{\psi_1 - V_1}{x_1} \tag{9.19}$$

and

$$E_2 = \frac{\psi_2 - V_2}{x_2} \tag{9.20}$$

where V_1 and V_2 are the bias voltage in n-Ge and p-GaAs respectively.

Substituting Equations (9.17) and (9.18) in Equations (9.19) and (9.20), we get

$$(\psi_1 - V_1)\varepsilon_{r1} \cdot N_d = (\psi_2 - V_2)\varepsilon_{r2} \cdot N_a \tag{9.21}$$

For heterojunction transistor from energy band diagram we see that electron current from n-Ge to p-GaAs is very small in comparison to the hole current which is flowing from opposite side because of the potential barrier of $(\psi_1 + \psi_2 + \Delta E_c/q)$ across the junction.

9.6.2 Application of heterojunction bipolar transistor (HBT)

HBTs are mainly used for high speed switching devices in the microwave frequency.

Problem 9.2 A heterojunction transistor at 300 K has the following parameters:

N_d for n-Ge = 5×10^{18} cm^{-3}
N_a for p-GaAs = 6×10^{16} cm^{-3}
Cross section (A) = 2×10^{-2} cm^2

Bias voltage at earlier junction (V'_E) = 1 V
Hole life time (τ_h) = 6×10^{-6} s
Calculate (a) the built-in voltage in the p-GaAs side, (b) the hole mobility, and (c) the hole diffusion constant.

Solution: (a) The built in voltage in the p-GaAs side is

$$\psi_2 = -26 \times 10^{-3} \ln \left(\frac{6 \times 10^{16}}{1.8 \times 10^{6}}\right) = -0.63 \text{ V}$$

(b) At this voltage the corresponding hole mobility obtained from the Appendix A is

$$\mu_h = 400 \text{ cm}^2/\text{V·s}$$

(c) The hole diffusion constant is

$$D_n = 400 \times 26 \times 10^{-3}$$
$$= 10.40 \text{ cm}^2/\text{s}.$$

REFERENCES

1. Sobol, H. and Sterzer, F., Solid state microwave power sources, *IEEE spectrum*, **4**, 32, April 1972.
2. Johnson, E.O., Physical limitations on frequency and power parameters of transistors, *RCA Rev.*, **26**, No. 6, 163–177, June 1965.

SUGGESTED READINGS

1. Liao, S.Y., *Microwave Devices and Circuits*, 3rd ed., Prentice-Hall of India, New Delhi, 1995.
2. Sze, S.M., *Semiconductor Devices of Physics and Technology*, 2nd ed., Wiley, New York, 2002.
3. Navon, D.H., *Semiconductor Microdevices and Materials*, Holt, Rinehart and Winston, New York, 1986.

PROBLEMS

1. Why cannot ordinary Bipolar Transistors be used in the high frequency range?
2. Mention the factors on which the maximum frequency of operation of a Bipolar Junction Transistor depends.
3. Explain the current-frequency, power-frequency and gain-frequency limitations of a high frequency Bipolar Junction Transistor.
4. Mention some of the applications of a Microwave Bipolar Transistor.
5. Calculate the maximum allowable power through silicon microwave BJT whose parameters are given below:

   Reactance (X_c) = 2 Ω
   Transit time cut-off frequency (f_T) = 6 GHz
   Maximum Electric field (E_m) = 10^5 V/cm
   Saturation drift velocity (v_s) = 2×10^5 cm/s

6. A Ge–GaAs heterojunction transistor has the following parameters:
   Lattice constant:
   Ge = 5.646 A
   GaAs = 5.653 A
   Electron affinity:
   Ge = χ_1 = 4 eV
   GaAs = χ_2 = 4. 02 eV
   Energy gap:
   Ge = 0.8 eV
   GaAs = 1.45 eV

   Determine

   I. The lattice match in percentage
   II. The conduction band differential between Ge and GaAs.
   III. The valence band differential between Ge and GaAs.

7. An n-Ge-p-GaAs-n-GaAs heterojunction transistor at 300 K has the following parameters:
   Donor density in the n-Ge region = 5×10^{18} cm^{-3}
   Acceptor density in the p-GaAs region = 3×10^{16} cm^{-3}
   Hole life time (τ_h) = 2×10^6 s
   Bias voltage across n-p junction (V_E) = 0.8 eV

Calculate:

(a) The built in voltage in the p-GaAs side.
(b) The hole mobility.
(c) The hole diffusion contact.
(d) The minority hole density in the p-GaAs region.
(e) The minority electron density in the p-GaAs region.
(f) The hole diffusion length.
(g) The emitter junction current.

8. Name some applicantion of HBT.

9. With the help of energy band diagram explain the operation of a heterojunction transistor.

10. In which way a heterojunction transistor is superior to a homojunction transistor?

Chapter 10

Metal Semiconductor Field Effect Transistors (MESFETs)

10.0 INTRODUCTION

In 1938 Schottky [1] introduced the idea of formation of a potential barrier due to the difference of work function between metal and semiconductor contacts. The same idea was utilized during the construction of metal semiconductor field effect transistor (MESFET). Mead [2] proposed MESFET in 1966. In 1969, Middlehoek [3] realized a silicon metal-semiconductor field effect transistor with a 1 μm gate length, which had a maximum frequency of oscillation of 12 GHz. In 1971 a significant step was made by Turner [4], when 1 μm gate length FETs on GaAs were made with f_{max} equal to 50 GHz. GaAs MESFET has higher electron mobility, higher electric field and higher electron saturation drift velocity than silicon devices, so its output power is also greater. Also, low noise figure of GaAs MESFET made it useful in microwave integrated circuits for high power, low noise and broadband amplifier applications. The operation of MESFET is similar to that of JFET, but the only difference is in MESFET, a metal semiconductor rectifying contact is used at the gate instead of a p-n junction of a JFET. Besides silicon and GaAs, InP can also be used as a substrate for MESFET fabrication. Drift velocity of InP is 50% higher than that of GaAs, which makes f_c, 1.6 times higher than that of GaAs [5]. Thus MESFET requires a semiconductor having large mobility, large maximum drift velocity and a large avalanche breakdown field. Materials, which are closer to the ideal in this respect, is either GaAs or InP.

10.1 STRUCTURE OF MESFET

A simple schematic diagram of MESFET on semi-insulating GaAs substrate is shown in Figure 10.1. By ion-implantation process dominant-impurity chromium (Cr) is doped into the GaAs substrate to bring the Fermi-level near the centre of the bandgap and results in a very high resistivity substrate of 100 μm thick. On this non-conducting substrate a thin layer of lightly doped (of about 0.15–0.35 μm thick) n-type GaAs is grown epitaxially to form the channel region of the FET. Sometimes the high resistivity (10^{15} to 10^{16} atoms/cc) GaAs buffer layer (≅ 3 μm) is grown between n-type GaAs layer and the substrate. By

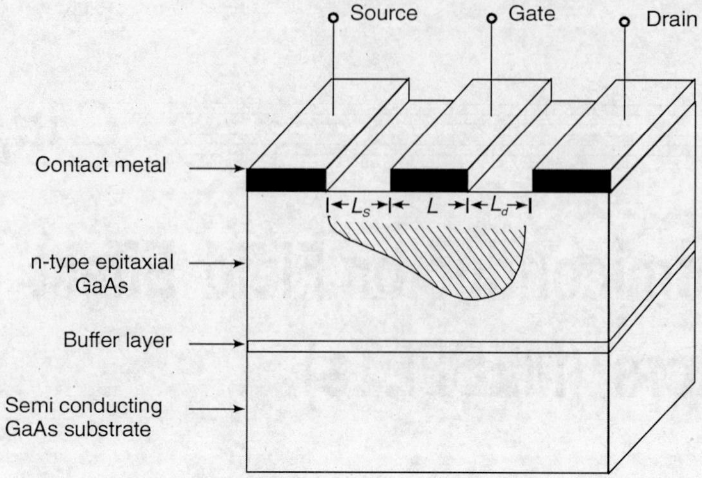

Figure 10.1 Schematic diagram of a GaAs MESFET.

photolithographic process metal layer of Au-Ge is deposited on an n-type GaAs for source and drain ohmic contacts and aluminium layer is used for Schottky barrier gate contact.

10.2 DEVICE FABRICATION AND PACKAGING

The fabrication of MESFET includes three major steps: (i) the growth and characterization of the semi-insulating substrate; (ii) the growth and characterization of the active-device region, which actually consists of one or more thin n-type epitaxial GaAs layers; and (iii) device packaging and testing.

The active region of the device can be grown up epitaxially by any one of the epitaxial growth techniques like MBE, VPE or MOCVD, over the substrate. After metallization the device is ready for packaging. The package should have minimal effect on the overall device performance. A typical package for low noise GaAs FET and the symbol is as shown in Figure 10.2(a) and (b).

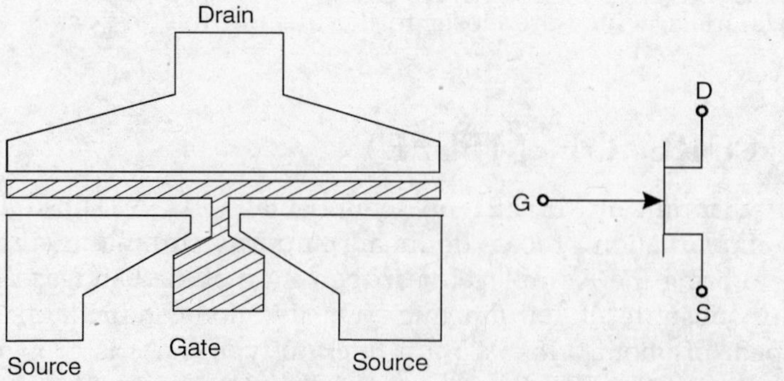

(a) MESFET geometry (b) Symbol of MESFET

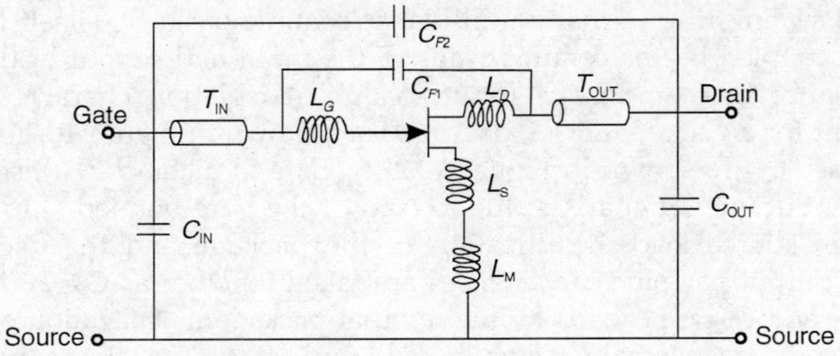

(c)

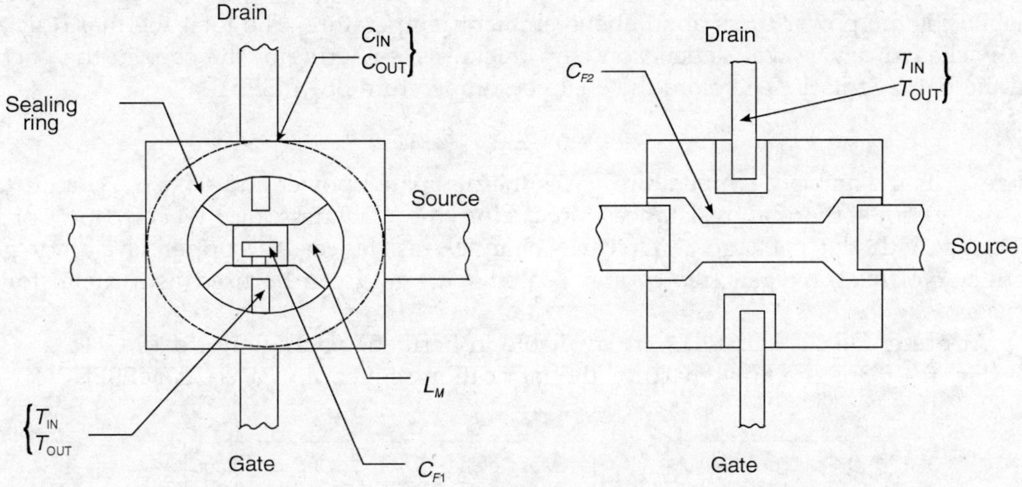

(d)

Figure 10.2 (a) MESFET geometry, (b) symbol of MESFET, (c) equivalent circuit for a GaAs MESFET, and (d) origins of package parasitics.

Package styles, materials, properties and manufacturing processes all influence their design. Such designs in turn influence device performance whilst assembly and sealing techniques affect performance, yield, cost and reliability. In case of GaAs MESFET the packages may be:

1. Confired ceramic type
2. Hard fired ceramic type
3. Glass-metal type

A confired multi-layer ceramic package involves the lamination and metallization of two or more layers of the confired ceramic and subsequent firing of the structure gives the final form of the package. This package is suitable for the operation of MESFET up to 18 GHz. Packages of the confired type are usually sealed with a gold plated thin metal lid using a gold tin solder preform.

Because of the high cost in comparison with the confired ceramic package, a glass metal package is not so much popular as a MESFET device package.

The metal sealing ring of many multi-layer ceramic packages, itself forms a resonant structure when coupled by the ceramic walls to the input and output leads. This produces a low Q resonance in the package input to output isolation characteristics. A typical equivalent circuit for a GaAs MESFET package is as shown in Figure 10.2(c) and the origins of package parasitics are shown in Figure 10.2(d). Gate inductance (L_G), source inductance (L_S) and drain inductance (L_D) arises due to lead of the wire bonded chip. The inductance (L_M) arises from source lead metallization of the package and the package mounting. Different input output parasitic capacitances are taken together as C_{IN} and C_{OUT}. Feedback capacitances C_{F1} and C_{F2} are caused by the internal package metallization geometry and the external lead sizes respectively. Finally T_{IN} and T_{OUT} are the components of the transmission lines. Now lower the value of these inductances and capacitances, higher will be the frequency of operation. Another important package property relating to device reliability is the power dissipation above ambient temperature. The total thermal resistance of a packaged device will depend on chip thickness, soldering of the chip in the package and the package itself. Therefore the channel temperature of MESFET

$$T_{CH} = T_A + P_D\,(\theta_{\text{chip}} + \theta_{\text{solder}} + \theta_{\text{mount}})$$

where T_A is the ambient temperature; P_D is the dissipated power and θ_{chip}, θ_{solder} and θ_{mount} are the various thermal resistances. Reducing the thickness of the chip and proper soldering with the package can reduce channel resistance. As copper is a very good conductor of heat, oxygen free copper coated with gold can be used as a metal for the package.

At present GaAs MESFETs are available in hermetic package up to 10 GHz.

Figure 10.3 is the flow chart for the fabrication sequence of a GaAs MESFET.

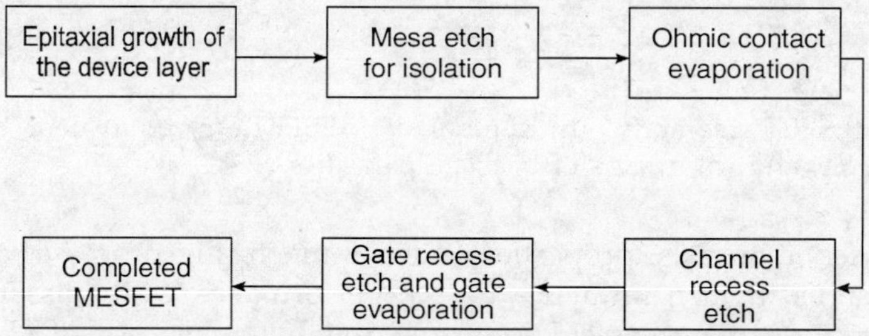

Figure 10.3 Fabrication sequence of GaAs MESFET.

10.3 MATERIALS FOR MESFETs

Apart from Si and GaAs, InP has been investigated as a substrate for MESFET fabrication. As InP has 50% higher drift velocity than GaAs; thus f_c for InP MESFET is much higher than that of GaAs MESFET, but the maximum frequency of oscillation ($f_{\max}$) is 20% lower. Also degenerate feedback resulting from a larger gate to drain capacitance and a smaller output resistance degrades the overall microwave performance.

MESFET requires a semiconductor of large mobility and large avalanche breakdown field, which results in a large maximum drift velocity. It means that the carriers must have a small effective mass and the materials must have a large inter-valley separation as well as a large energy gap. Materials, which are suitable for this other than GaAs, InP are the so called quaternary crystals such as InAs-InP and InAs-GaAs [6, 7].

10.4 PRINCIPLE OF OPERATION OF MESFET

A positive voltage V_{DS} between drain and source causes electrons to flow from drain to source end of the device. For small values of V_{DS} the layer appears to be a linear resistor. The current flows through the channel so that the Schottky barrier gate electrode becomes progressively more reverse biased with respect to the drain electrode, that causes a voltage drop along the length of the channel. As a result of this charge depletion is set up in the channel and gradually pinches off the channel against the semi-insulating substrate towards the drain end. With the increase of reverse bias between source and the gate the width of the charge depletion layer increases and results in a decrease in channel width. This decrease in channel width increases the channel resistance and which in turn decreases the rate of increase of current. Thus the drain current I_D will be modulated by the gate voltage $V_{GS.}$ Figure 10.4 below shows a family of curves of the drain current I_D versus the drain to source voltage V_{DS} with the gate voltage V_{GS} as one of the parameters and can be drawn for a unipolar Schottky-barrier gate of a GaAs MESFET.

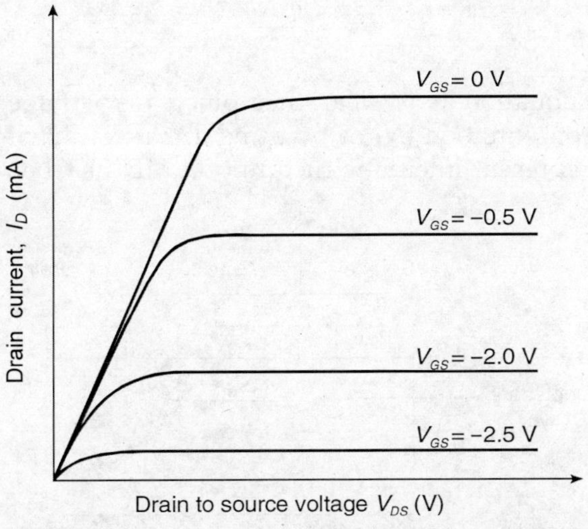

Figure 10.4 Voltage-current characteristics of n-channel MESFET.

As the drain current (I_D) is controlled by the field effect of the gate voltage (V_{GS}), the device is called the Field Effect Transistor (FET). As mentioned earlier because of the high reverse bias voltage between the gate and the drain, the channel is eventually pinched off. At pinched off, the drain current (I_D) remains almost constant even if the drain to source voltage (V_{DS}) is continuously increased.

The pinch off voltage is the gate reverse voltage that removes all the free charge from the channel. Poisson's equation in the n-channel in terms of volume charge density is

$$\frac{d^2V}{dy^2} = -\frac{\rho}{\varepsilon} = \frac{-q_e N}{\varepsilon_r \varepsilon_0} \tag{10.1}$$

where N is the electron concentration, q_e is the charge of the electron; and ε_r, ε_0, the permittivity of the material and free space respectively.

If a be the length of the channel, then using the boundary condition

$$E = -\frac{dV}{dy} = 0 \quad \text{at} \quad y = a$$

or

$$\frac{dV}{dy} = -\frac{q_e N}{\varepsilon}(y - a) \text{ V/m} \tag{10.2}$$

Again, integrating within the boundary condition at $y = 0$, $V = 0$ results in

$$V = -\frac{q_e N}{2\varepsilon}(y^2 - 2ay) \tag{10.3}$$

using Equation (10.3), the pinch off voltage (V_p) is at $y = a$.

At $y = a$, when $V = V_p$, pinch off takes place and the pinch off voltage is given by

$$V_p = \frac{q_e N a^2}{2\varepsilon} \tag{10.4}$$

Thus from the above equation it is clear that pinch off voltage can be increased by increasing the doping concentration (N) or by using a material having low permittivity (ε). Figure 10.5(a) and (b) represent internal structure of a MESFET before pinch off and after pinch off.

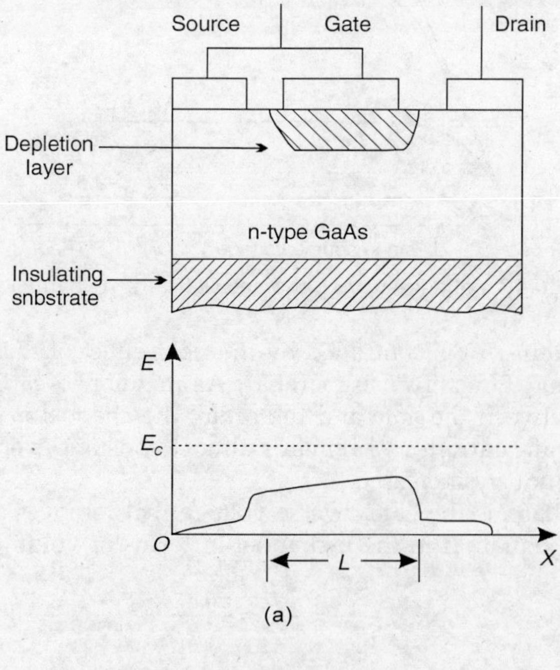

(a)

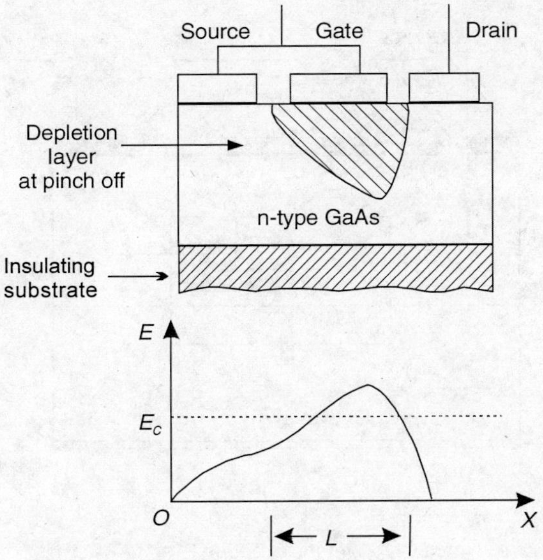

(b)

Figure 10.5 (a) Depletion layer and electric field before pinch off and (b) depletion layer and electric field at pinch off.

In practice, the saturation drain current depends on V_{GS} and can be approximated by

$$I_D = I_{DS}\left(1 - \frac{V_{GS}}{V_p}\right)^2 \tag{10.5}$$

and $I_D = I_{DS}$ for $V_{DS} > V_{DS,\text{sat}}$. I_{DS} is equal to I_D at $V_{GS} = 0$. From Equation (10.5) the transconductance g_m (Siemens) can be defined as

$$g_m = \left(\frac{dI_D}{dV_{GS}}\right)_{V_{DS}=\text{constant}}$$

or

$$g_m = -\frac{2I_{DS}}{V_p}\left(1 - \frac{V_{GS}}{V_p}\right) \tag{10.6}$$

10.5 SMALL SIGNAL EQUIVALENT CIRCUIT

The small signal equivalent circuit is useful for circuit design at a low power level. Figure 10.6(b) is the small signal equivalent circuit and the physical origin of these components is as shown in Figure 10.6(a).

(a) The various components in the model are defined as the following:

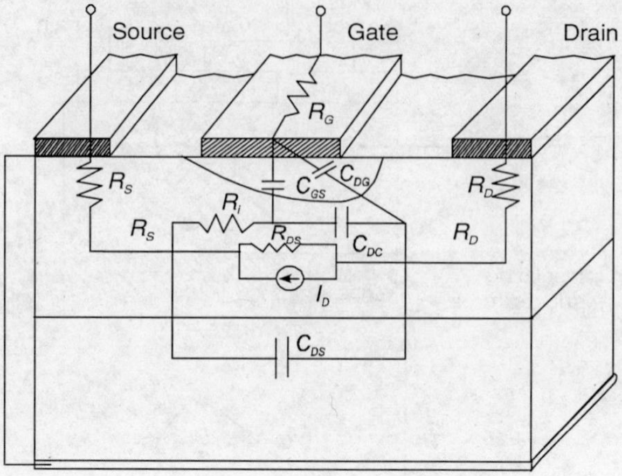

(a)

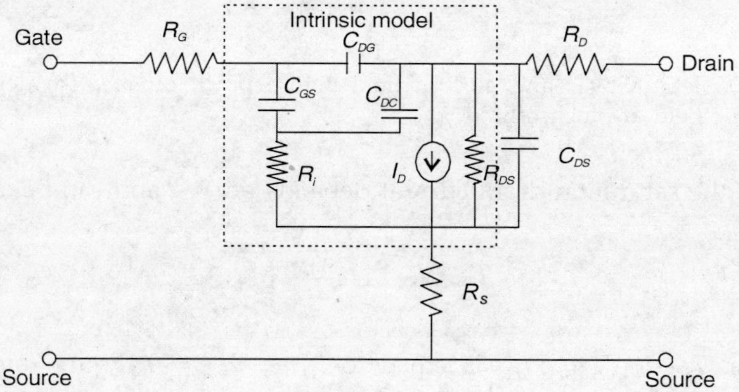

(b)

Figure 10.6 (a) Physical origin of the circuit elements of an MESFET and (b) small signal equivalent circuit of an MESFET.

INTRINSIC ELEMENTS:
R_i : Input (channel) resistance
C_{GS} : Gate to source capacitance
C_{DG} : Gate to drain feedback capacitance
R_{DS} : Drain to source resistance
C_{DC} : Drain to channel capacitance

EXTRINSIC ELEMENTS:
C_{DS} : Drain to source capacitance
R_D : Drain to channel resistance, including contact resistance
R_S : Source to channel resistance, including contact resistance
R_G : Gate metal resistance

Parasitic inductances L_G, L_S and L_D, due to the lead could be added in series with R_G, R_S and R_D in the equivalent circuit. The values of these elements depend on the channel doping, channel type, material and dimensions. The large values of the extrinsic resistance reduce the power gain and efficiency significantly, and increase the noise figure. The feedback capacitance increasing the channel doping (N) can reduce C_{GD} and which also increases the transconductance (g_m) and the dc open circuit voltage gain. Again increase of N decreases the breakdown voltage of the gate.

As derived by Zuleeg and Lehovec [8] the expression for drain current I_D:

$$I_D = I_{DS} \frac{3[(\mu^2 - \rho^2) - 2(\mu^3 - \rho^3)]}{1 + \eta(\mu^2 - \rho^2)} A \tag{10.7}$$

where I_{DS} is the saturation current at $V_{GS} = 0$

$$= \frac{q_e N a \mu z V_p}{3L}$$

q_e = Charge of an electron
a = Length of the channel
z = Channel depth
L = Gate length
V_p = Pinch off voltage
μ = Mobility of electron (for n-type channel) in $m^2/V{\cdot}s$

$u = \left(\frac{V_{DS} + |V_{GS}|}{V_p}\right)^{1/2}$ is the normalized value of the drain and gate voltage w.r.t. the pinch off voltage.

$$\rho = \left(\frac{|V_{GS}|}{V_p}\right)^{1/2}$$

$$\eta = \frac{\mu |V_p|}{v_s L} = \frac{|v|}{v_s} = \frac{\text{normalized drift velocity}}{\text{saturated drift velocity}}$$

where

$v = \dfrac{\mu E_x}{1 + \dfrac{\mu E_x}{v_s}}$ is the drift velocity in the channel for electric field (E_x) of the channel.

The drain current is maximum at $u = u_m$

Then
$$u_m^3 + 3u_m\left(\frac{1}{z} - \rho^2\right) + 2\rho^3 - \frac{3}{z} = 0 \tag{10.8}$$

where $z = \mu |V_p| / (v_x L)$

Substituting Equation (10.8) into Equation (10.7), we get

$$I_D = 3I_{DS}\,(1 - u_m)/z \tag{10.9}$$

The transconductance in saturation region is given by:

$$g_m = \left.\frac{dI_D}{dV_{GS}}\right|_{V_{DS}=\text{constant}} = \frac{g_{\max}(u_m - \rho)}{1 + z(u_m^2 - \rho^2)} \tag{10.10}$$

Ohkawa, et al. [9] has shown that the equivalent circuit shown in Figure 10.6 has a critical frequency f_c above which the MESFET is unconditionally stable. f_c can be approximated as

$$f_c = \frac{1}{2\pi(\tau_0 + \tau_1 + \tau_2)} \tag{10.11}$$

The transmittance $Y_m = g_m e^{-j\omega\tau_0}$

$$\tau_1 = \frac{C_{DG}(2R_G + R_i + R_s)}{\left[\dfrac{C_{DG}}{C_{GS}} + \dfrac{R_s}{R_{DS}}\right]} \tag{10.12}$$

and

$$\tau_2 = \frac{2}{\dfrac{g_m}{C_{GS}}\left[\dfrac{C_{DG}}{C_{DS}} + \dfrac{R_s}{R_{DS}}\right]} \frac{R_G + R_i + R_s}{R_{DS}} \tag{10.13}$$

Below this cut-off frequency the device becomes unstable since larger fraction of the output voltage is fed back to the input,

where $$R_{\text{in}} = \frac{1}{2\omega^2 C_{GS}^2 (R_G + R_i + R_s)}$$

Thus

$$R_{\text{in}} \propto \frac{1}{\omega^2} \text{ and the feedback capacitance } C_{DG} \propto \frac{1}{\omega}$$

The device unilateral power gain can be defined as

$$G_{\max} = (f_{\max}/f_c)^2 \tag{10.14}$$

Thus the maximum frequency of oscillation

$$f_{\max} = \frac{f_c}{2\sqrt{r_1 + f_c\tau_3}}$$

where $$r_1 = \frac{R_G + R_i + R_s}{R_{DS}}$$

and $$\tau_3 = 2\pi R_G C_{DG}$$

alongwith R_G and R_s, the feedback capacitance (C_{DG}) must be minimized.

The maximum available gain at a frequency f of the MESFET is given by

$$G_{\max} = \left(\frac{f_c}{f}\right)^2 \left[\frac{4}{R_{DS}}(R_i + R_s + R_G + \omega_c L_s)\right]^{-1} + 2\omega_c\, C_{DG}\, (R_i + R_S + 2R_G + \omega_c L_s) \tag{10.15}$$

Where L_s is the inductance in the common source lead of the device. The experimental value for the maximum frequency of oscillation for GaAs MESFET with the gate length less than 10 micron is.

$$f_{\max} \cong \frac{33 \times 10^3}{L}\ \text{Hz} \tag{10.16}$$

where L is the active length of the channel in micron.

The maximum values of $f_{\max}$ is obtained by the transit time τ as mentioned in Equation (10.11), it is used to find out the cut-off frequency f_c.

Therefore

$$f_c = \frac{1}{2\pi\tau} = \frac{v_s}{2\pi L} = \frac{g_m}{2\pi C_{GS}}\ \text{Hz} \tag{10.17}$$

The highest frequency of oscillation for maximum power gain given by Dingle, et al. [10] for matched network is

$$f_{\max} = \frac{f_c}{2}\left[\frac{R_D}{R_s + R_G + R_i}\right]^{1/2}\ \text{Hz} \tag{10.18}$$

Problem 10.1 A GaAs MESFET has the following parameters
Electron concentration (N): = 7×10^{17} cm^{-3}
Electronic charge (q_e) = 1.6×10^{-19} coulomb
Relative permittivity (ε_r) = 11
Channel length (a) = 0.1 µm
Calculate the pinch off voltage.

Solution From Equation (10.4) the pinch off voltage

$$V_p = \frac{q_e N a^2}{2\varepsilon_0 \varepsilon_r} = \frac{1.6 \times 10^{-19} \times 7 \times 10^{17} \times (0.1 \times 10^{-6})^2}{2 \times 8.854 \times 10^{-12} \times 11}$$
$$= 0.005\ \text{mV}.$$

Problem 10.2 A GaAs MESFET has the following parameters $R_G = 6\ \Omega$, $R_i = 2\ \Omega$

$g_m = 60$ m℧, $R_s = 2\ \Omega$, $R_{DS} = 400\ \Omega$, $C_{GS} = 0.5$ pF and $C_{DG} = 0.1$ pF.

Calculate (i) the cut-off frequency, (ii) maximum operating frequency.

Solution: (i) From Equation (10.17) the cut-off frequency

$$f_c = \frac{g_m}{2\pi C_{GS}} = \frac{60 \times 10^{-3}}{2 \times 3.14 \times 0.5 \times 10^{-12}} = 19.1\ \text{GHz}$$

(ii) Maximum frequency of oscillation

$$f_{\max} = \frac{f_c}{2\sqrt{r_1 + f_c \tau_3}}$$

$$\text{Now } r_1 = \frac{(R_G + R_i + R_s)}{R_{DS}} = \frac{6 + 2 + 2}{400} = \frac{1}{40} = 0.025$$

$$\tau_3 = 2\pi R_G C_{DG} = 2 \times 3.14 \times 6 \times 0.1 \times 10^{-12}$$

$$= 3.768 \times 10^{-12} \text{ s.}$$

Therefore,

$$f_{\max} = \frac{19.1 \times 10^9}{2\sqrt{0.025 + 19.1 \times 10^9 \times 3.768 \times 10^{-12}}}$$

$$= \frac{9.55 \times 10^9}{0.31} = 30.8 \text{ GHz.}$$

Problem 10.3 Calculate (i) the pinch off voltage (V_p), (ii) the velocity ratio (η), (iii) saturation current at V_{GS} = 0, (iv) the drain current (I_D) of an n-channel GaAs MESFET, having the following parameters:

Permittivity of free space (ε_0)	= 8.854×10^{-12} F/m.
Channel width (a)	= 0.15 μm
Electron concentration (N)	= 8×10^{23} m^{-3}
Channel length (L)	= 10 μm
Breadth of the channel (z)	= 30 μm
Electron mobility (μ)	= 800 cm^2/V·s
Drain voltage (V_{DS})	= 5 V
Gate voltage (V_{GS})	= – 2.0 V
Saturation drift velocity (v_s)	= 2×10^5 m/s
Relative dielectric constant (ε_r)	= 11.2

Solution: (i) From Equation (10.4) the expression for pinch off voltage

$$V_p = \frac{q_e N a^2}{2\varepsilon} = \frac{q_e N a^2}{2\varepsilon_0 \varepsilon_r}$$

$$= \frac{1.6 \times 10^{-19} \times 8 \times 10^{23} \left(0.15 \times 10^{-6}\right)^2}{2 \times 8.854 \times 10^{-12} \times 11.2} = 1.452 \text{ V}$$

(ii) Velocity ratio (η) can be expressed as

$$\eta = \frac{\mu |V_p|}{v_s L} = \frac{0.08 \times 1.452}{2 \times 10^5 \times 10 \times 10^{-6}} = 0.058$$

(iii) The expression for drain to source current at V_{GS} = 0 is

$$I_{DS} = \frac{q_e N \mu a z V_p}{3L}$$

$$= \frac{1.6 \times 10^{-19} \times 8 \times 10^{23} \times 0.08 \times (0.15 \times 10^{-6}) \times 30 \times 10^{-6} \times 1.452}{3 \times 10 \times 10^{-6}}$$

$$= 0.167 \times 10^{-2}\,\text{A}$$

$$= 1.67\,\text{mA}$$

(iv) Again we know that

$$u = \left(\frac{V_{DS} + |V_{GS}|}{V_p}\right)^{1/2} = \left(\frac{5+2}{1.452}\right)^{1/2} = 2.19$$

and

$$\rho = \left(\frac{|V_{GS}|}{V_p}\right)^{1/2} = \left(\frac{2}{1.452}\right) = 1.173$$

$$I_D = I_{DS} \cdot \frac{3(u^2 - \rho^2) - 2(u^3 - \rho^3)}{1 + \eta(u^2 - \rho^2)}\,\text{A}$$

$$= 1.67 \times 10^{-3} \times \frac{|3\{(2.19)^2 - (1.173)^2\} - 2\{(2.19)^3 - (1.173)^3\}|}{1 + 0.058\{(2.19)^2 - (1.173)^2\}}$$

$$= \frac{1.67 \times 10^{-3} \times [3(4.7961 - 1.376) - 2(10.503 - 1.613)]}{1 + 0.058(4.7961 - 1.376)}$$

$$= \frac{1.67 \times 10^{-3}(10.26 - 17.78)}{1 + (0.058 \times 3.42)} = 4.2091\,\text{A}$$

10.6 DUAL GATE MESFET

The two independent gates increase the capability of MESFET in terms of improvement of signal gain and it reduces feedback as well as gain control and mixing. The dual gate MESFET can be used in modulators, up-down converters, phase shifters, discriminators and also as an oscillator (*rf* source). In 1971, Turner et al. [11] first fabricated dual gate MESFET and its analysis was done by Liechti [12] in 1975. The equivalent circuit and cross-sectional view of dual gate MESFET are as shown in Figure 10.7(a) and (b). From the structure it is clear that a dual gate MESFET is two separate MESFETs in cascade. Here the top device determines the characteristic of the bottom MESFET. Two MESFETs when combined, dual gate MESFET is obtained. The analytical model of the dual gate MESFET used by Statz et al. [13] is as shown in Figure 10.8. Using this analytical model, one can easily calculate the drain currents, (I_{D_1} and I_{D_2}), the channel lengths of a dual gate MESFET. Thus by adjusting

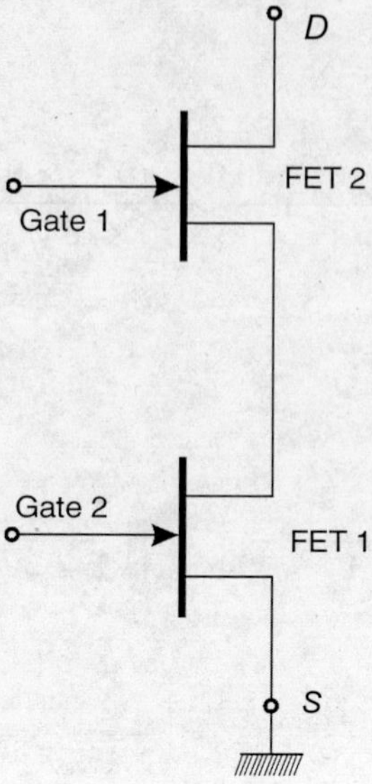

(a)

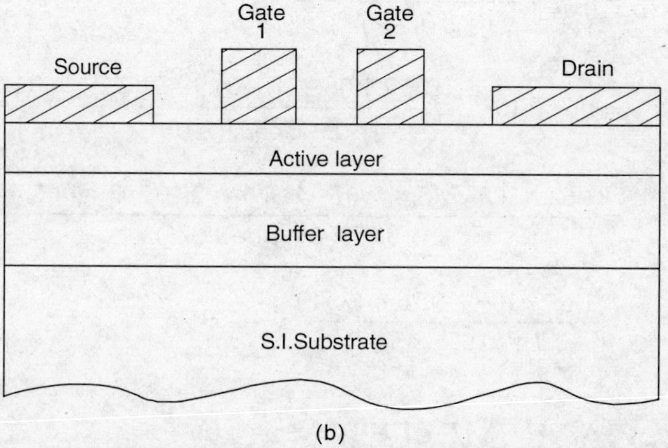

(b)

Figure 10.7 (a) Equivalent circuit of dual gate MESFET and (b) cross sectional view of dual gate MESFET.

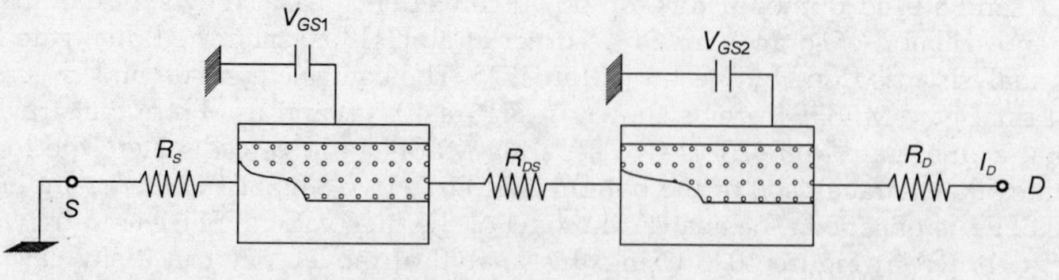

Figure 10.8 Analytical model of a dual gate MESFET.

$I_{D_1} = I_{D_2} = I_D$ for given bias condition V_{GS_1}, V_{GS_2} and V_{DS} the calculated and measured V–I characteristic of a MESFET is as shown in Figure 10.9.

The dual gate MESFET is generally characterized as a 3-port device with the source connected to ground. Here the transmission coefficient is high between gate 1 and drain and also considerable transmission from gate 1 to gate 2. The forward transmission

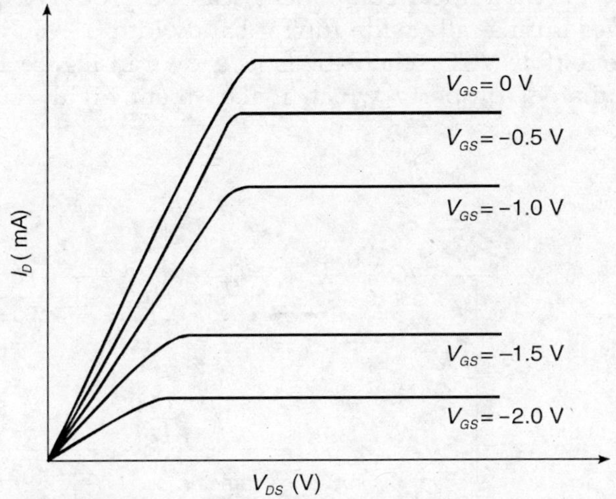

Figure 10.9 Voltage current characteristic of a dual gate MESFET.

coefficient between gate 2 and drain is lower than gate 1 to drain because the output impedance of MESFET-1, $1/g_{d_1} + R_s + \omega L_{s_1}$ is in series with the second gate capacitance C_{gs_2} and the integrate resistance R_{IG}, thus lowering the transconductance of the second gate. The output impedance of the first MESFET also acts as series feedback impedance increasing the reverse transmission coefficient from drain to gate 2. The simplified equivalent circuit of a dual gate MESFET is as shown in Figure 10.10. The equivalent circuit is drawn on the basis of cascading of two MESFETs and can be used safely up to 12 GHz.

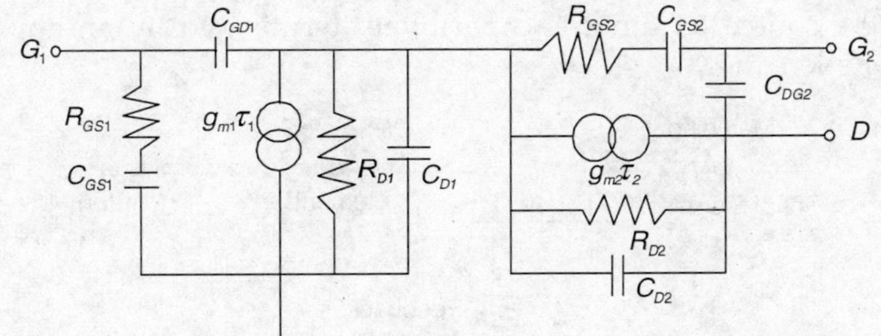

Figure 10.10 Simplified equivalant circuit of a dual gate MESFET.

10.7 MESFET OSCILLATORS

MESFET oscillators are available up to a frequency of 25 GHz. Like other microwave devices in MESFET negative resistance cannot be achieved by application of dc bias only. The condition has therefore to be induced in these devices by using proper circuit topology so that useful power at microwave frequencies can be obtained.

The first work on microwave GaAs MESFET oscillators in which a packaged device with inductive impedance at the gate and capacitive inductance at the source was reported by Maeda et al. [14]. These days MM1C voltage controlled oscillator (VCO) with MESFET is very popular because all the critical components can be produced photolithographically and this VCO can offer intrinsically wide tuning bandwidth.

A simplified monolithic VCO schematic is as shown in Figure 10.11. GaAs dual gate MESFETs have an inherent property which makes them highly suitable for low chirp

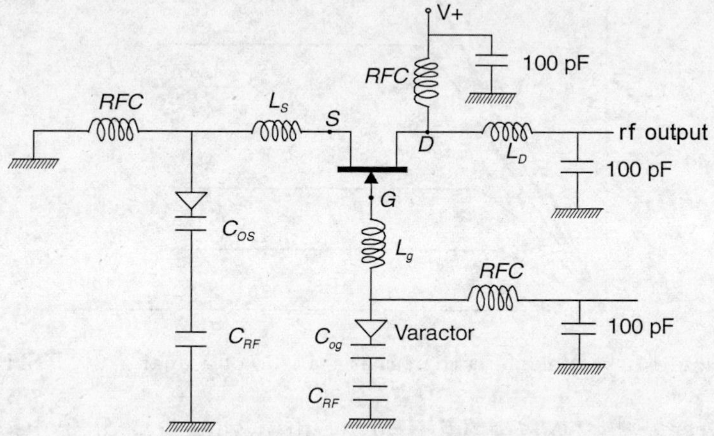

Figure 10.11 Schematic diagram of VCO using MESFET.

pulsed rf oscillator applications. If the first MESFET containing gate 1 is used as an active element for obtaining steady state oscillation, the second MESFET containing gate 2 can be operated as a high-speed switch to obtain pulsed rf output. By applying a negative pulse to the second gate, this can be done easily. Figure 10.12 shows a schematic diagram of a dual gate GaAs MESFET pulsed oscillator. A series feedback element is introduced at the gate 1 terminal while the source terminal is capacitively terminated by an appropriate 50 ohm short-circuited line length.

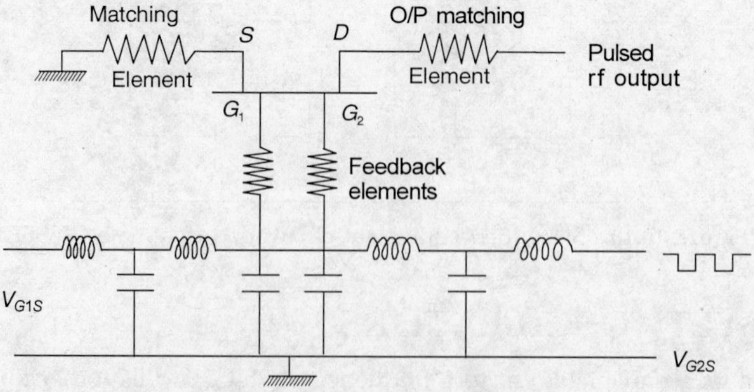

Figure 10.12 Diagram of dual gate MESFET pulsed oscillator.

10.8 MESFET AMPLIFIERS

Though MESFET can operate above 20 GHz, it is very difficult to design low noise MESFET amplifier over 20 GHz, because of the relatively low associated gain of the first few stages of a multistage amplifier. As mentioned earlier, a GaAs MESFET can be operated as a negative resistance two terminal device by inducing the correct impedance with feedback. A negative resistance amplifier using common drain MESFET is as shown in Figure 10.13.

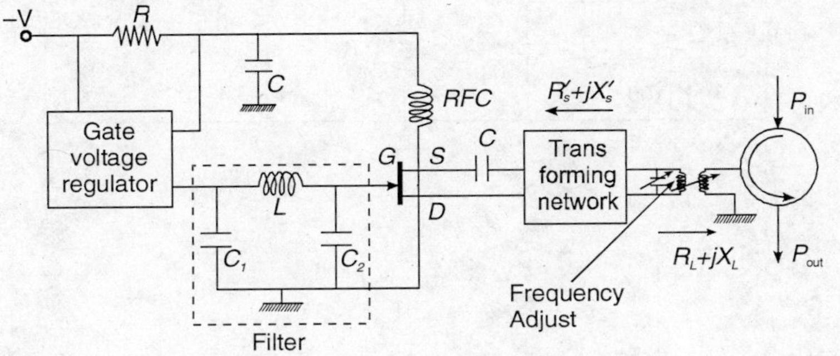

Figure 10.13 Negative resistance amplifier using common drain MESFET.

The condition for delivering maximum power to the load:

$$X_L = -X'_s$$

$$R_L < |R'_s|$$

and gain

$$G = \frac{P_{out}}{P_{in}} = \left|\frac{(R'_s - R_L)}{(R'_s + R_L)}\right|^2$$

A circulator is used to provide isolation between the input and output signal. At a frequency lower than 4 GHz, the reactance associated with the parasitic of MESFETs is small enough to enable its use without the need for matching components. In MESFET amplifier circuit of Figure 10.14(a) three MESFETs are used—one in common source and the other two in source-follower mode to produce a 50 ohm output impedance with stable operation. This circuit has a bandwidth of 1 dB and approximate frequency of operation is 1.5 GHz and which can be extended to 3 GHz by introducing some inductive peaking into the drain load of the input stage. Figure 10.14(b) shows the gain frequency response. Figures 10.15(a) and 10.15(b) show how inductive peaking at the interstage between the common source and source follower MESFET improves the bandwidth over 5 GHz with an increase in gain of 5.5 dB. The presence of the shunt 50 ohm resistor at the input of amplifier will increase the noise figure.

For a power GaAs MESFET, larger gate widths are required and the structure efficiency has to be achieved with minimum effect of parasitic elements. Both horizontal and vertical channels can be used. Vertical channel geometry has a better power frequency response but suffer from higher parasitic gate source and gate drain capacitance.

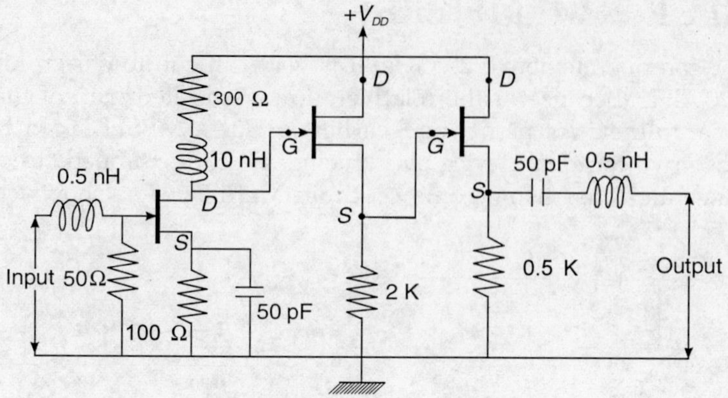

(a)

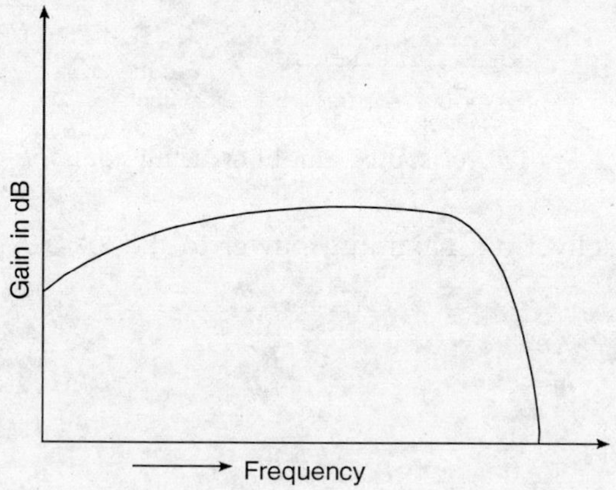

(b)

Figure 10.14 (a) Low frequency MESFET amplifier circuit with inductive peaking and (b) gain frequency response of common source MESFET amplifier.

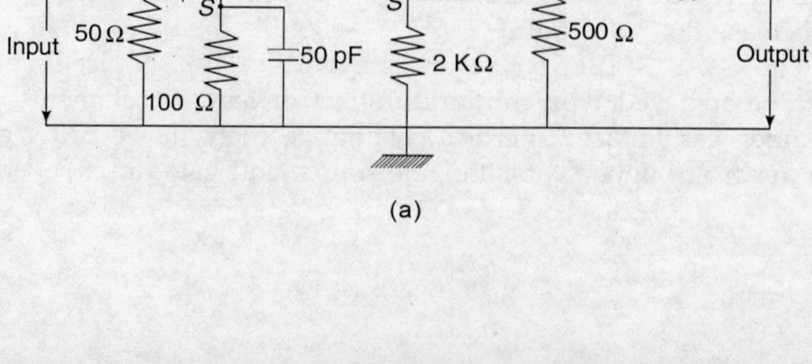

(a)

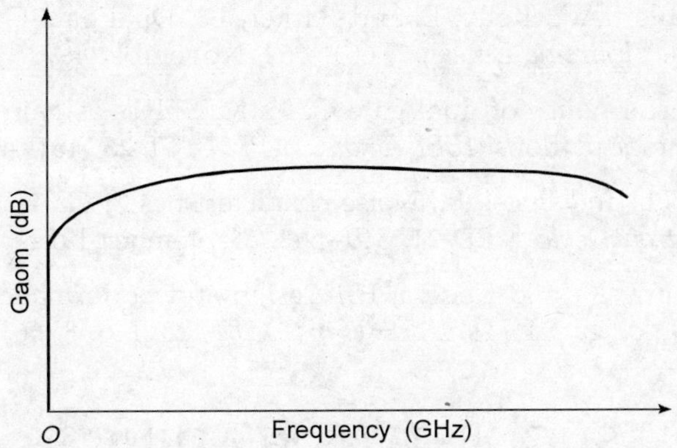

Figure 10.15 (a) Low frequency MESFET amplifier with interstage inductive peaking and (b) gain frequency response.

REFERENCES

1. Schottky, W., *Naturwiss;* **26**, 843 (1938).
2. Mead, C.A., Schottky barrier gate field effect transistor, *Proc. IEEE-ED*, **2**, 307–308, February 1966.
3. Middlehoek, S., Projection masking, thin photoresist layers and interface effects, *IBM J. Res. Develop*, **14**, 117–124, March 1970.
4. Turner, J.A., Waller, A., Bennet and Parker, D., An electron beam fabricated GaAs microwave field effect transistor, *Symposium on* GaAs *and Related Compounds, Inst. Physics. Conf.*, Serial No. 9, 234–239, 1971.
5. Barreara, J. and Archer, R., InP Schottky gate field effect transistors, *IEEE Trans Electron Devices*, **ED-22**, 1023–1030, November 1973.
6. Fawcett, W., Hilsum, C., and Ress, H., Optimum semiconductor for microwave devices, *Electronics Letters*, **5**, 313–314, July 1969.
7. Glicksman, M., Enstrom, R., Mittleman, S. and Appert, J., Electron mobility in $In_x Ga_{1-x}$ As alloys, *Phys. Rev. B.*, **9**, 1621–1626; February 1974.
8. Zuleeg, R. and Lehovec, K., High frequency and temperature characteristics of GaAs Junction Field effect Transistors in the hot electron range, *Proc. Symp.* GaAs, *Institute of Physics Conf.*, Series No. 9, 240–245, 1970.
9. Ohkawa, S., Soyama, K., and Ishikawa, H., Low noise GaAs field effect transistor, *Fujitsu Sci Tech. J.*, Vol. II, 151–173, March 1975.
10. Dingle, R., et al., Electron mobilities in modulation–doped semi-conductor hetero-junction superlattices, *Applied Phys. Letters,* Vol. 33, 665–667, 1978.

11. Turner, J.A., Waller, A.J., Kelly, E. and Parker, D., Dual gate GaAs microwave field effect transistors, *Electronic Letters*, **7**, 661–662, November 1971

12. Liechti, C.A., Performance of dual gate GaAs MESFETs as gain controlled amplifiers and high speed modulations, *IEEE Trans. MTT*, **MTT-23**, 461–469, June 1975.

13. Statz, H., Hans, H., and Pucel, R., Noise characteristics of GaAs field effect transistors, *IEEE Trans. Electron Devices*, **ED-21**, 549–562 , September 1974.

14. Maeda, M., Kimura, K., and Kodera, H., Design and performance of X-band oscillator with GaAs Schottky gate FETs, *IEEE Trans. MTT*, **23**, No. 8, 661–667, August 1975.

SUGGESTED READINGS

1. Raymond, S., Pengelly, *Microwave Field-Effect Transistors—Theory, Design and Applications*, 2nd ed., Research Studies Press, John Wiley & Sons, New York, 1986.

2. Sisodia, M.L. and Gupta, V.L., *Microwaves—Introduction to Circuits, Devices and Antenna*, New Age International Publisher, New Delhi, 2001.

3. Liao, S.Y., *Microwave Devices and Circuits*, 3rd ed., Prentice-Hall of India, New Delhi, 1991.

4. Pulfrey, David L. and Garry Tarr, N., *Introduction to Microelectronic Devices*, Prentice Hall Inc., Englewood Cliffs, N.J., 1989.

PROBLEMS

1. What are MESFETs? Explain the construction, operation and performance characteristics and their applications.
2. Which parameters of a MESFET control its pinch off voltage?
3. Draw the small signal equivalent circuit of a MESFET and mention the sources of different components of MESFET equivalent circuit.
4. For a MESFET $g_m = 55$ ℧, $R_g = 3\ \Omega$, $L = 150\ \mu\text{m}$, $v_0 = 2 \times 10^7$ cm/s. Calculate f_c and f_{max}.
5. A GaAs MESFET has the following parameters $R_G = 6\ \Omega$, $R_i = 2\ \Omega$, $g_m = 60$ m℧, $R_s = 2\ \Omega$, $R_{DS} = 400\ \Omega$, $C_{GS} = 0.5$ pF and $C_{DG} = 0.1$ pF. Calculate (i) the cut-off frequency, (ii) the maximum operating frequency.
6. A typical GaAs MESFET has the following parameters:

Electron concentration (N)	$= 8 \times 10^{17}$ cm^{-3}
Channel height (a)	= 0.12 μm
Relative dielectric constant (ε_r)	= 12.5
Channel length (L)	= 15 μm
Channel width (z)	= 38 μm
Electron mobility (μ)	= 800 cm^2/V·s
Drain voltage (V_{DS})	= 5 V
Gate voltage (V_{GS})	= –2 V
Saturation drift velocity (v_s)	$= 2 \times 10^5$ m/s

Calculate (i) the pinch off voltage; (ii) the velocity ratio, (iii) saturation current at zero gate voltage, and (iv) the drain current.

7. A MESFET has a gate width of 10 μm. Calculate the maximum frequency of oscillation for the device.

8. A GaAs MESFET has a thickness of 0.40 μm and a doping concentration of 5×10^{17} cm^{-3}. The relative dielectric constant is 12.5. Calculate the pinch off voltage in volts of the device.

9. An n-channel GaAs MESFET has the following parameters

Electron concentration (*N*)	$= 2 \times 10^{18}$ cm^{-3}
Channel height (*a*)	= 0.2 μm
Relative dielectric constant (ε_r)	= 11.5
Channel length (*L*)	= 10 μm
Channel width (*z*)	= 60 μm
Electron mobility (μ)	= 0.3 m^2/V·s
Drain voltage (V_{DS})	= 5 V
Gate voltage (V_{GS})	= −2 V
Saturation drift velocity	= 10^5 m/s

Write a computer program to calculate the pinch off voltage, the velocity ratio and the drain current at zero gate voltage.

10. Mention some of the limitations of MESFET.

11. Mention some of the parameters, which limit the frequency of operation of a MESFET.

12. What is dual gate MESFET? What is its advantage over single gate MESFET?

13. Write down some of the applications of dual gate MESFET.

Chapter 11

High Electron Mobility Transistor (Modulation Doped Field Effect Transistor)

11.0 INTRODUCTION

HEMT (High Electron Mobility Transistor) also termed MODFET (Modulation Doped Field Effect Transistor) have emerged as highly suitable devices for low noise amplification of microwaves and millimeter waves. The basic principle of operation is same as Junction Field Effect Transistors (JFET). Ohmic contacts are provided at the source and drain terminals and a reverse biased depletion layer is formed at the gate electrode. Unlike JFET the gate electrode does not form a p-n junction with the substrate but a metal semiconductor Schottky barrier diode is formed at the gate terminal, which provides a depletion layer beneath the gate when the same is reverse biased with respect to the substrate (Figure 11.1).

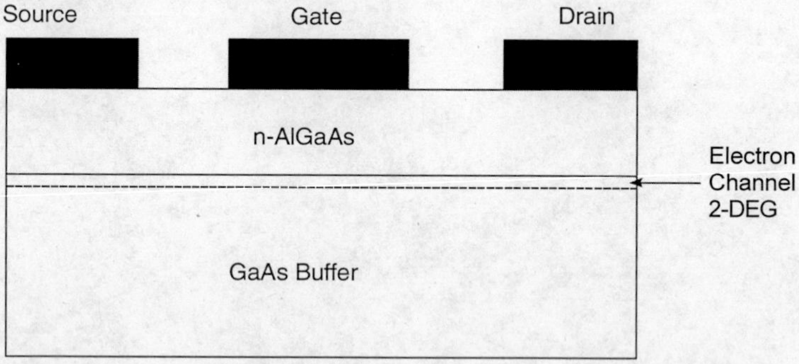

Figure 11.1 Simplified structure of the AlGaAs/GaAs HEMT.

Electrons flow from drain to source and the gate voltage controls the current through the channel cross-section by changing the extension of the depletion layer. In the case of Metal Semiconductor Field Effect Transistor, the substrate is a high resistivity semi insulating GaAs, which is formed by doping with silicon atoms, which bring the Fermi level to the centre of the band gap. The positioning of the Fermi level produces high resistivity GaAs crystal wafers called semi-insulating substrate. When the substrate is doped by ion-implantation a conducting n-GaAs substrate is formed. The conducting n-GaAs layer on the substrate can also be formed by growth of an epitaxial layer by molecular beam epitaxy or by MOVPE (Metallo Organic Vapour Phase Epitaxy). In MESFET this conducting n-layer forms the channel under the gate electrode. The gate electrode is an aluminium layer, which forms a Schottky diode with n-GaAs. When the gate electrode is reverse biased the current through the n-channel between the source and the drain is controlled by the variation of the gate voltage. The frequency of operation is dependent on the time the electrons take to pass beneath the gate to reach the source from the drain electrode. The gate length is therefore a critical parameter, which is made small for millimeter wave application.

For higher amplification factor and high speed [1] GaAs layer has to be made more conducting by increasing the doping level. Increasing the doping, however, leads to more ionized impurity scattering in n-GaAs, which causes lower mobility for the charge carriers. In high electron mobility transistors, a heterojunction of AlGaAs and GaAs is formed over the semi-insulating substrate and under the metal (aluminium) Schottky gate. On reverse biasing the electrons from the conduction band of AlGaAs enters the potential well at the heterojunction into the conduction band of GaAs. This leads to the creation of a sheet of high-density electrons in the conduction band of the doped GaAs which is termed as two dimensional electron-gas (2-DEG). Electron in 2-DEG can move very fast in GaAs, where there is no constraint of high-ionized impurity scattering. The high electron mobility in 2-DEG leads to very fast low noise operation in the devices. HEMT or MODFET are now used as low noise amplifiers well into mm waves up to 70 GHz, which has recently been developed up to 94 GHz (W-Band) by reducing the gate length. They have proved to be highly useful for X-Band, Ka-Band and Ku-Band amplification and are suitable not only as discrete but also in monolithic microwave integrated circuits. Reducing the temperature of the device to liquid nitrogen temperature (which leads to much reduced ionized impurity scattering and much larger electron mobility) can considerably enhance HEMT amplification. Such HEMT amplifiers would be suitable in radio astronomy as low noise receivers.

11.1 HETEROJUNCTION AND 2-DEG

HEMT has a heterojunction of AlGaAs and GaAs over a semi insulating substrate of GaAs. The heterojunction is formed at the interface of epitaxially grown layers of compound semiconductors, which have different band gap energy. Lattice constants of $Al_xGa_{1-x}As$ (where, x is the mole fraction, i.e., molecular proportion of Al to Ga) is nearly same as GaAs. These layers are therefore lattice matched and can be grown by Molecular Beam Epitaxy (MBE) technique or by MOCVD (Metallo-organic chemical vapour deposition) technique over semi-insulating GaAs substrate.

Consider the band diagram of a slightly doped layer of GaAs of bandgap 1.42 eV at 300 K (Figure 11.2) forming a heterojunction with a layer of highly doped n-AlGaAs having a different bandgap (2.1 eV) (Figure 11.3). Hence there are large number of mobile electrons in AlGaAs with a corresponding number of immobile positive doner ions. The electrons from the AlGaAs layer of the heterojunction will easily pass over the heterojunction and enter the conduction band of the GaAs and become trapped in the potential well. There the electrons in GaAs cannot surmount the potential barrier of the heterojunction but are attracted by positive doner ions at the interface of the heterojunction. The GaAs layer has thus a high density of electron, which forms a highly conducting channel between source

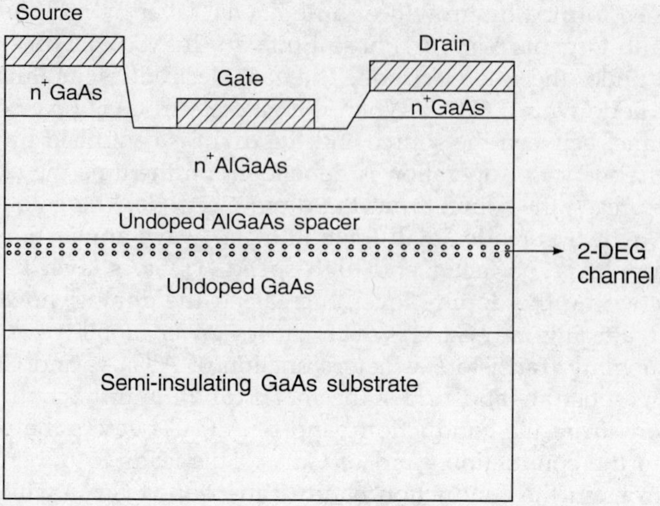

Figure 11.2 Actual structure of a AlGaAs/GaAs HEMT.

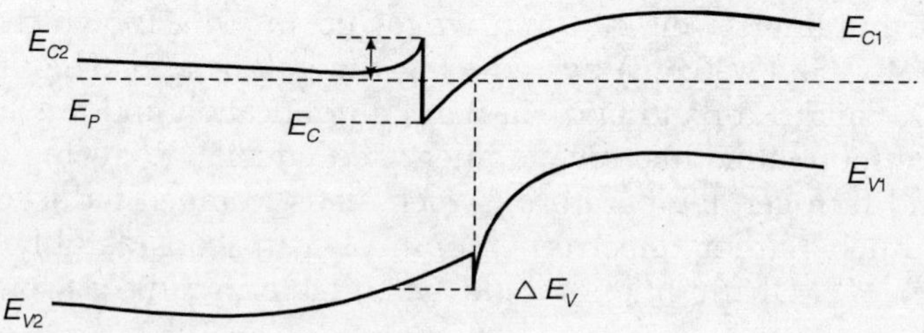

Figure 11.3 Band diagram of the heterojunction between a doped AlGaAs and a GaAs undoped buffer layer over a GaAs SI substrate.

and drain. These high-density electrons in GaAs are confined to the interface of the heterojunction due to the attraction of the positive donor ions of AlGaAs as shown in Figure 11.4; and form what is known as 2-dimensional electron gas (2-DEG). These electrons have entered from AlGaAs and have not been created by doping GaAs and they

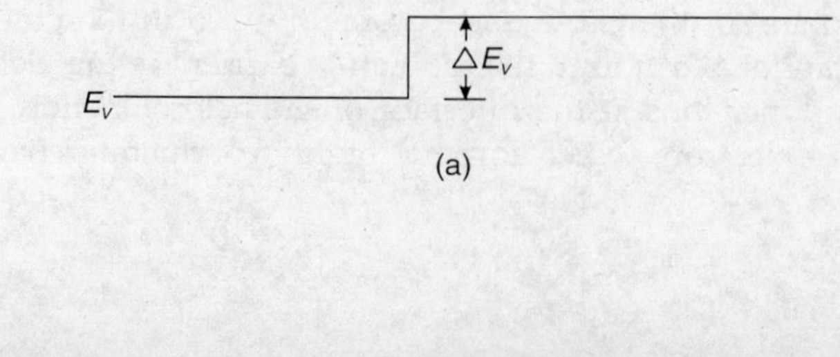

(a)

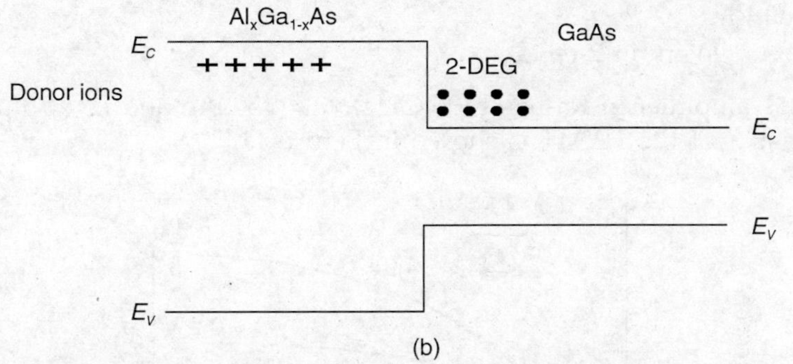

Figure 11.4 (a) Simplified energy band of the heterojunction between doped Al_xGa_{1-x} As and undoped GaAs and (b) After transfer of free electrons from AlGaAs to GaAs which is at the interface due to attraction of positive donor ions forming 2-DEG, i.e., a sheet of electrons.

cannot escape towards AlGaAs. They can only move in two dimensions parallel to the plane of the heterojunction. Hence the term 2-dimensional electron gas is applied to this sheet of highly conducting electrons. The mobility of the electrons is very high since the ionized impurity scattering does not decrease in drift velocity as in the case of highly doped GaAs. The mobility is high and is that in bulk GaAs with no impurities where the mobility is limited by Lattice scattering. At 300 K, the mobility is 8000 $cm^2/V{\cdot}s$, which increases to 250,000 $cm^2/V{\cdot}s$ at 77 K due to reduced lattice scattering and to 2×10^6 $cm^2/V{\cdot}s$ at 4 K.

11.2 DEVICE STRUCTURE AND FABRICATION

HEMT is based on a heterojunction between heavily doped n^+ AlGaAs and undoped GaAs. The 2-dimensional electron gas is formed below the interface and lies within the potential well created by conduction band discontinuity as discussed earlier. The cross section of HEMT is shown in Figure 11.2. The electrons in HEMT flow in undoped GaAs. Ionized impurity scattering is thus reduced and the carriers have high mobility and drift velocity. The undoped AlGaAs spacer layer improves the mobility by further separating 2-DEG from the donor ions in AlGaAs. This layer is kept very thin (< 50 Å).

HEMT layers are grown by Molecular Beam Epitaxy (MBE) because of the requirements of thin layers and abrupt transitions. They have also been grown by MOCVD (Metallo Organic Chemical Vapour Deposition). The short gates (0.2 – 0.3 μm) for mm wave HEMTs are patterned by electron beam lithography.

11.3 HEMT CHARACTERISTICS

The device current in a HEMT is dependent on the gate width w (perpendicular to the gate length) that is much larger than the gate length. The drain current in HEMT is given by

$$I_D = qnwv$$

where

q = Electronic charge
n = Electron concentration in the 2-dimensional gas

w = Gate width
v = Electron velocity in 2-DEG.

For a HEMT amplifier in Ka-Band (36 GHz), w = 150 μm and L = 0.5 μm. A schematic of *I-V* characteristics of the HEMT is shown in Figure 11.5.

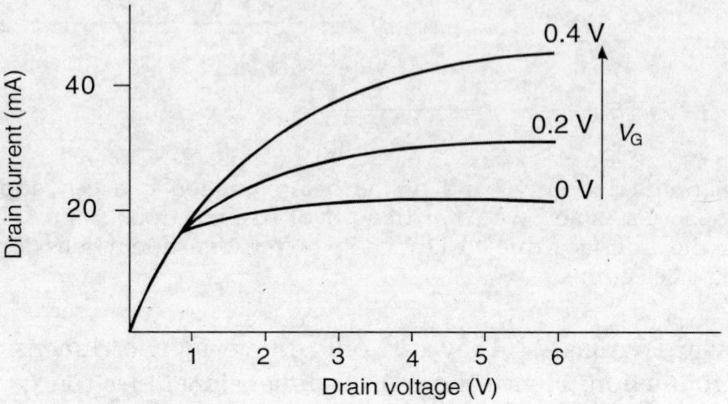

Figure 11.5 I_D–V_{DS} characteristics against gate voltage.

The equivalent circuit of a HEMT is shown in Figure 11.6 where the parameters depend on the operating frequency band.

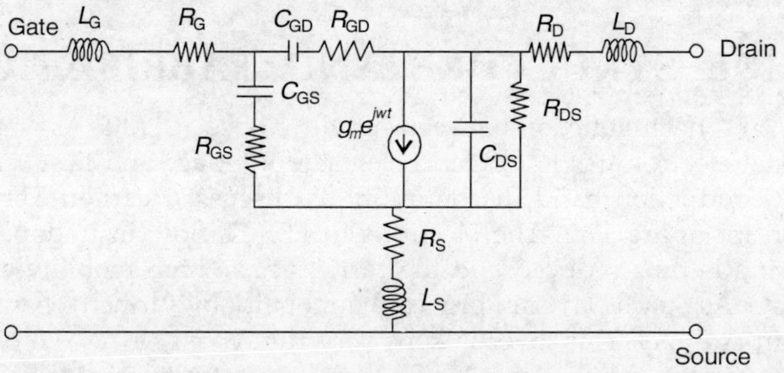

Figure 11.6 High frequency equivalent circuit of a HEMT.

The rate of change of threshold voltage with the thickness of $Al_xGa_{1-x}As$ layer is called threshold sensitivity of the device and is given by

$$\frac{dV_{th}}{dw} = -[2N_d q(\psi_{ms} - \Delta E_C - V_{th})/\varepsilon]^{1/2}$$

where

q = Electronic charge
N_d = Donor concentration
V_{th} = Threshold voltage
w = Thickness of AlGaAs layer
ψ_{ms} = Metal-semiconductor Schottky barrier

E_C = Conduction band edge difference between GaAs and AlGaAs
ε = Permittivity of AlGaAs
= Free space permittivity × dielectric constant of AlGaAs.

The maximum possible frequency of operation of a HEMT is determined by the transit time required by electrons to cross the gate while flowing from drain to source. The cut-off frequency is given by

$$f_c = \frac{1}{2\pi\tau_{\text{transit}}} = \frac{v}{2\pi L}$$

where
v = Carrier velocity, and
L = Gate length

Thus extremely small values of L (between 0.1 μm and 500 nm) are necessary for HEMTs to operate between 60–94 GHz, which can be realized by electron-beam photolithography, and high quality masks.

HEMTs have very high transconductance. Typical values for 0.25 μm HEMTs range between 400–600 ms/nm. The high transconductance and the low input capacitance for short gate length lead to good high frequency characteristics. The large gain bandwidth product of the HEMT makes it suitable for broadband amplification. HEMT has also shown gain of 9.1 and 5.3 dB at 60–94 GHz.

The noise figure of HEMT is lower than FET due to very low ionized impurity scattering. Presently 0.25 μm HEMT exhibits noise figures of 1.7, 2.6 and 3.7 dB at 30, 40 and 44 GHz (see Figure 11.7).

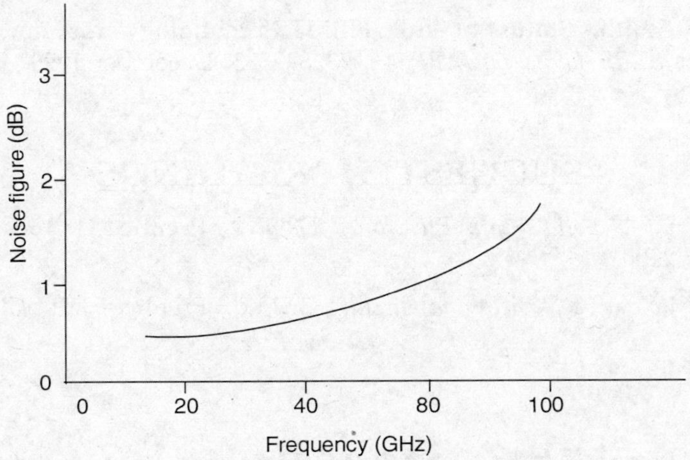

Figure 11.7 Noise figure of a typical InP HEMT against frequency.

11.4 RECENT DEVELOPMENT OF HEMT

Although most microwave HEMTs are based on GaAs/AlGaAs systems, it has been shown in recent years that InP based HEMT having structure, InAlAs/$In_xGa_{1-x}As$/InAlAs/InP have higher f_{max} than a GaAs based HEMT and have superior noise performance. Large band discontinuity at the InGaAs/InAlAs heterojunction permits high 2-dimensional electron gas (2-DEG) densities to be obtained resulting in high current and high transconductance. This in turn leads to increase in operating frequency and higher gain bandwidth. Typical 0.1 μm InP low noise HEMT exhibit g_m of 800-1000 ms/nm compared with 600 ms/nm for comparable GaAs HEMTs. The current status of InP based HEMT technology indicates [2] that low noise amplification for receivers can be realized up to 100 GHz.

11.5 CONCLUSION

HEMT transistors have lowest noise figure at frequencies ranging from 1–60 GHz. At cryogenic temperature, noise figure and bandwidth show great improvement making this suitable for receivers in Radio Astronomy. HEMT is extremely suitable for broadband applications and is having an increasing number of space applications as satellite transmitters and receivers. HEMTs are three terminal devices, hence they are suitable for integration in MMIC circuits and HEMT integrated MMIC with necessary impedance matching are having increasing applications as microwave and millimeter wave amplifiers.

REFERENCES

1. Abe, M., Recent advances in ultra high speed HEMT technology, *IEE Trans. on Quantum Electronics*, **QE 22,** 1870–1879, September 1986.
2. Phillip M. Smith, Status of InP HEMT Technology for microwave Receiver Applications, *IEEE Trans. on MTT*, **44,** 2328–2333, December 1996.

SUGGESTED READINGS

1. Streetman, B.G., *Solid State Electronic Devices*, Prentice-Hall of India, 2nd ed., New Delhi, 2000.
2. Liao, S.Y., *Microwave Devices and Circuits*, 3rd ed., Prentice-Hall of India, New Delhi, 2003.

PROBLEMS

1. Calculate the drain current of a HEMT having the following parameters:
   Gate width: 120 μm
   Electron velocity: 3.5×10^5 m/s
   2D electron gas density: 5×10^{15} m^{-2}

2. Calculate the conduction band edge difference between GaAs and AlGaAs and the sensitivity of a HEMT having the following parameters:
Threshold voltage (V_{th}): 14 V
Donor concentration (N_d): 2.5×10^{24} m^{-3}
Metal-semiconductor Schottky Barrier potential (ψ_{ms}): 0.8 V
AlGaAs dielectric constant (ε_r): 4.43
GaAs bandgap (E_{gg}): 1.43 V
AlGaAs bandgap (E_{ga}): 1.80 V

3. Describe the structure and fabrication procedure of HEMT.

4. Explain the operating principle of HEMT and also mention some of its uses.

5. Mention some of the constructional differences of HEMT from that of a MESFET.

6. Explain the formation of 2-DEG at the interface of the heterojunction of a HEMT.

Appendix

Table I Physical Constants

Velocity of light, $c = 2.998 \times 10^8$ m/s
Charge of electron, $e = 1.602 \times 10^{-19}$ C
Mass of electron, $m = 9.107 \times 10^{-31}$ kg
Charge to mass ratio of an electron, $e/m = 1.76 \times 10^{11}$ C/kg
Boltzmann constant, $k = 1.380 \times 10^{-23}$ J/K
Planck's constant, $h = 6.547 \times 10^{-34}$ J·s
Permittivity of vacuum, $\varepsilon = 8.854 \times 10^{-12} = (1/36\pi) \times 10^{-9}$ F/m
Permeability of vacuum, $\mu_0 = 4\pi \times 10^{-7}$ H/m
Impedance of free space, $Z_0 = 376.7 = 120\ \pi\Omega$

Table II Characteristics of Single Stage Semiconductor Microwave Amplifiers

Name of the device	*Gain* (dB)	*Centre frequency* (GHz)	*Noise figure*
BJT	4–10	1–10	2–6
FET	3–30	10–28	2–5
Tunnel diode	5–15	4–15	5
TED	7–10	5–37	15
IMPATT diode	5–10	6–10	20

Table III Dielectric Constant—Relative Permittivity (ε_r)

Material	ε_r
Air	1
Alcohol (ethyl)	25
Bakelite	4.8
Glass	4–7
Ice	4.2
Mica (ruby)	5.4
Nylon	4

Table III Dielectric Constant—Relative Permittivity (ε_r) (cont.)

Material	ε_r
Paper	2–4
Plexiglass	2.6–3.5
Polyethylene	2.25
Polystyrene	2.55
Porcelain (dry process)	6
Quartz (fused)	3.8
Rubber	2.5–4
Sands (dry)	4
Silica (fused)	3.8
Snow	3.3
Sodium chloride	5.9
Soil (dry)	2.8
Styrofoam	1.03
Teflon	2.1
Water (distilled)	80
Water (sea)	20
Water (dehydrated)	1
Wood (dry)	1.5–4
Ground (wet)	5–30
Ground (dry)	2–5
Water (fresh)	80

Table IV Relative Permeability, μ_r

Diamagnetic material	μ_r	*Ferromagnetic material*	μ_r
Bismuth	0.99999860	Nickel	50
Paraffin	0.99999942	Cast iron	60
Wood	0.99999950	Cobalt	60
Silver	0.99999981	Machine steel	300
		Ferrite (typical)	1,000
Paramagnetic material	μ_r	Transformer iron	3,000
Aluminium	1.00000065	Silicon iron	4,000
Beryllium	1.00000079	Iron (pure)	4,000
Nickel chloride	1.00004	Mumetal	20,000
Manganese sulphate	1.0001	Supermalloy	100,000

Table V Properties of Semiconductor Materials

	E_g (eV)	μ_n ($cm^2/V{\cdot}s$)	μ_p ($cm^2/V{\cdot}s$)	ρ (Ω-cm)	*Transition*	*Doping*	*Lattice*	q (Å)	ε_r	*Density* (g/cm^3)	*Melting point* (°C)
Si	1.11	1350	480	2.5×10^5	*i*	n,p	*D*	5.43	11.8	2.33	1415
Ge	0.67	3900	1900	43	*i*	n,p	*D*	5.66	16.	5.32	936
SiC(α)	2.86	500		10^{10}	*i*	n,p	*W*	3.08	10.2	3.21	2830
AlP	2.45	80		10^{-5}	*i*	n,p	*Z*	5.46		2.40	2000
AlAs	2.16	180		0.1	*i*	n,p	*Z*	5.66	10.9	3.60	1740
AlSb	1.6	200	300	5	*i*	n,p	*Z*	6.14	11	4.26	1080
GaP	2.26	300	150	1	*i*	n,p	*Z*	5.45	11.1	4.13	1467
GaAs	1.43	8500	400	4×10^8	*d*	n,p	*Z*	5.65	13.2	5.31	1238
GaSb	0.7	5000	1000	0.04	*d*	n,p	*Z*	6.09	15.7	5.61	712
InP	1.28	4000	100	8×10^{-3}	*d*	n,p	*Z*	5.87	12.4	4.79	1070
InAs	0.36	22,600	200	0.03	*d*	n,p	*Z*	6.06	14.6	5.67	943
InSb	0.18	10^5	1700	0.06	*d*	n,p	*Z*	6.48	17.7	5.78	525
ZnS	3.6	110		10^{10}	*d*	n	*Z,W*	5.409	8.9	4.09	1650
ZnSe	2.7	600		10^9	*d*	n	*Z*	5.671	9.2	5.65	1100
ZnTe	2.25		100	100	*d*	p	*Z*	6.101	10.4	5.51	1238
CdS	2.42	250	15		*d*	n	*W,Z*	4.137	8.9	4.82	1475
CdSe	1.73	650		10^5	*d*	n	*W*	4.30	10.2	5.81	1258
CdTe	1.58	1050	100	10^{10}	*d*	n,p	*Z*	6.482	10.2	6.20	1098
PbS	0.37	575	200	5×10^{-3}	*i*	n,p	*H*	5.936	161	7.6	1119
PbSe	0.27	1000	1000	10^{-3}	*i*	n,p	*H*	6.147	280	8.73	1081
PbTe	0.29	1600	700	10^{-2}	*i*	n,p	*H*	6.452	360	8.16	925

Table VI Energy Gap Structures of Semiconductors

Semi-conductor	*Energy gap* (eV)	*Lattice constants* (Å)	*Energy gap structure*	*Expansion coefficient at* 300°K ($\times 10^{-6}$ °C^{-1})	*Hetero-junction preferred doping*	*Typical dopants*	*Electron affinity* (eV)
$Ge_{0.9}Si_{0.1}$	0.77	(5.63)	Indirect	—	n	P, As, Sb	(4.1)
Ge	0.66	5.658	Indirect	5.7	p	Al, Ga, In	4.13
GaAs	1.43	5.654	Direct	5.8	n	Se, Te	4.07
Ge	0.66	5.658	Indirect	5.7	p	Al, Ga, In	4.13
ZnSe	2.67	5.667	Direct	7.0	n	Al, Ga, In	4.09
Ge	0.66	5.658	Indirect	5.7	p	Al, Ga, In	4.13
ZnSe	2.67	5.667	Direct	7.0	n	Al, Ga, In	4.09
GaAs	1.43	5.654	Direct	5.8	p	Zn, Cd	4.07
AlAs	2.15	5.661	Indirect	5.2	p	Zn	3.5
GaAs	1.43	5.654	Direct	5.8	n	Se, Te	4.07
GaP	2.25	5.451	Indirect	5.3	n	Se, Te	4.3
Si	1.11	5.431	Indirect	2.33	p	Al, Ga, In	4.01
AlSb	1.6	6.136	Indirect	3.7	n/p	Se, Te/Zn, Cd	3.65
GaSb	0.68	6.095	Direct	6.9	p/n	Zn, Cd/Se, Te	4.06
GaSb	0.68	6.095	Direct	6.9	n	Se, Te	4.06
InAs	0.36	6.058	Direct	4.5(5.3)	p	Zn, Cd	4.9
ZnTe	2.26	6.103	Direct	8.2	p	Cu	3.5
GaSb	0.68	6.095	Direct	6.9	n	Se, Te	4.06
ZnTe	2.26	6.103	Direct	8.2	p	Cu	3.5
InAs	0.36	6.058	Direct	4.5(5.3)	n	Se, Te	4.9
ZnTe	2.26	6.103	Direct	8.2	p	Cu	3.5
AlSb	1.6	6.136	Indirect	3.7	n	Se, Te	3.65
CdTe	1.44	6.477	Direct	5	p/n	Li, Sb, P/I	4.28
PbTe	0.29	6.52	Indirect	19.8	n/p	Cl, Br/Na, K	4.6
CdTe	1.44	6.477	Direct	5	p	Li, Sb	4.28
InSb	0.17	6.479	Direct	4.9	n	Se, Te	4.59
ZnTe	2.26	6.103	Direct	8.2	p	Cu	3.5
CeSe(hex)	1.7	4.3($\sqrt{2}$) (6.05)	Direct	2.45–4.4	n	Cl, Br, I	4.95

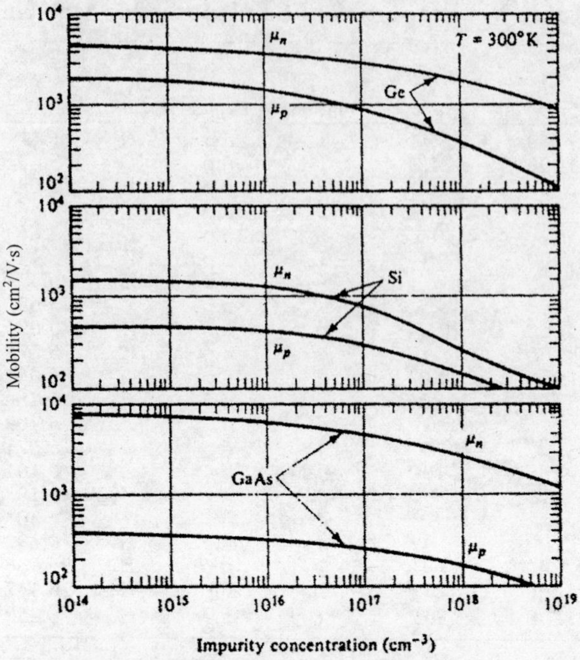

Figure I Mobility for Ge, Si, and GaAs.

Table VII Conductivity σ in mhos per metre

Conductor	σ	*Insulator*	σ
Silver	6.17×10^7	Quartz	10^{-17}
Copper	5.80×10^7	Polystyrene	10^{-16}
Gold	4.10×10^7	Rubber (hard)	10^{-15}
Aluminium	3.82×10^7	Mica	10^{-14}
Tungsten	1.82×10^7	Porcelain	10^{-13}
Zinc	1.67×10^7	Diamond	10^{-13}
Brass	1.50×10^7	Glass	10^{-12}
Nickel	1.45×10^7	Bakelite	10^{-9}
Iron	1.03×10^7	Marble	10^{-8}
Bronze	1.00×10^7	Soil (sandy)	10^{-5}
Solder	0.70×10^7	Sands (dry)	2×10^{-4}
Steel (stainless)	0.11×10^7	Clay	10^{-4}
Nichrome	0.10×10^7	Ground (dry)	10^{-4}– 10^{-5}
Graphite	7.00×10^4	Ground (wet)	10^{-2}– 10^{-3}
Silicon	1.20×10^3	Water (distilled)	2×10^{-4}
Water (sea)	3–5	Water (fresh)	10^{-3}
		Ferrite (typical)	10^{-2}

Index